Algebra 1

LARSON
BOSWELL
KANOLD
STIFF

Applications • Equations • Graphs

Chapter 4
Resource Book

The Resource Book contains the wide variety
of blackline masters available for Chapter 4.
The blacklines are organized by lesson. Included
are support materials for the teacher as well as
practice, activities, applications, and assessment
resources.

 McDougal Littell
A HOUGHTON MIFFLIN COMPANY
Evanston, Illinois • Boston • Dallas

Contributing Authors

The authors wish to thank the following individuals for their contributions to the Chapter 4 Resource Book.

Rita Browning
Linda E. Byrom
José Castro
Rebecca S. Glus
Christine A. Hoover
Carolyn Huzinec
Karen Ostaffe
Jessica Pflueger
Barbara L. Power
Joanne Ricci
James G. Rutkowski
Michelle Strager

ISBN: 0-618-02042-X

12 13 14 15 -CKI- 07 06 05

Contents

4 Graphing Linear Equations and Functions

Contents

Contents

Descriptions of Resources

This Chapter Resource Book is organized by lessons within the chapter in order to make your planning easier. The following materials are provided:

Tips for New Teachers These teaching notes provide both new and experienced teachers with useful teaching tips for each lesson, including tips about common errors and inclusion.

Parent Guide for Student Success This guide helps parents contribute to student success by providing an overview of the chapter along with questions and activities for parents and students to work on together.

Prerequisite Skills Review Worked-out examples are provided to review the prerequisite skills highlighted on the Study Guide page at the beginning of the chapter. Additional practice is included with each worked-out example.

Strategies for Reading Mathematics The first page teaches reading strategies to be applied to the current chapter and to later chapters. The second page is a visual glossary of key vocabulary.

Lesson Plans and Lesson Plans for Block Scheduling This planning template helps teachers select the materials they will use to teach each lesson from among the variety of materials available for the lesson. The block-scheduling version provides additional information about pacing.

Warm-Up Exercises and Daily Homework Quiz The warm-ups cover prerequisite skills that help prepare students for a given lesson. The quiz assesses students on the content of the previous lesson. (Transparencies also available)

Activity Support Masters These blackline masters make it easier for students to record their work on selected activities in the Student Edition.

Alternative Lesson Openers An engaging alternative for starting each lesson is provided from among these four types: *Application, Activity, Graphing Calculator,* or *Visual Approach.* (Color transparencies also available)

Graphing Calculator Activities with Keystrokes Keystrokes for four models of calculators are provided for each Technology Activity in the Student Edition, along with alternative Graphing Calculator Activities to begin selected lessons.

Practice A, B, and C These exercises offer additional practice for the material in each lesson, including application problems. There are three levels of practice for each lesson: A (basic), B (average), and C (advanced).

Contents

Reteaching with Additional Practice These two pages provide additional instruction, worked-out examples, and practice exercises covering the key concepts and vocabulary in each lesson.

Quick Catch-Up for Absent Students This handy form makes it easy for teachers to let students who have been absent know what to do for homework and which activities or examples were covered in class.

Cooperative Learning Activities These enrichment activities apply the math taught in the lesson in an interesting way that lends itself to group work.

Interdisciplinary Applications/Real-Life Applications Students apply the mathematics covered in each lesson to solve an interesting interdisciplinary or real-life problem.

Math and History Applications This worksheet expands upon the Math and History feature in the Student Edition.

Challenge: Skills and Applications Teachers can use these exercises to enrich or extend each lesson.

Quizzes The quizzes can be used to assess student progress on two or three lessons.

Chapter Review Games and Activities This worksheet offers fun practice at the end of the chapter and provides an alternative way to review the chapter content in preparation for the Chapter Test.

Chapter Tests A, B, and C These are tests that cover the most important skills taught in the chapter. There are three levels of test: A (basic), B (average), and C (advanced).

SAT/ACT Chapter Test This test also covers the most important skills taught in the chapter, but questions are in multiple-choice and quantitative-comparison format. (See *Alternative Assessment* for multi-step problems.)

Alternative Assessment with Rubrics and Math Journal A journal exercise has students write about the mathematics in the chapter. A multi-step problem has students apply a variety of skills from the chapter and explain their reasoning. Solutions and a 4-point rubric are included.

Project with Rubric The project allows students to delve more deeply into a problem that applies the mathematics of the chapter. Teacher's notes and a 4-point rubric are included.

Cumulative Review These practice pages help students maintain skills from the current chapter and preceding chapters.

LESSON 4.1

INCLUSION Students with limited English proficiency may have trouble learning all the vocabulary introduced on page 203. To help these students, introduce one term at a time using a drawing of a coordinate plane. You may wish to label the axes, origin, and quadrants as shown on page 203. You can help these students by making a connection between the coordinate axis and the real number line introduced in Lesson 2.1. Students can transfer the skill of locating points on the number line to ploting points in the coordinate plane.

COMMON ERROR After completing Example 2 on page 204, some students might think that all given data points must perfectly fit on the line of best fit of a scatter plot. Make sure to complete an exercise such as Example 3, where the line of best fit does not go through all data points. Many students believe that the line of best fit goes through the leftmost and right-most points of their graph, ignoring all the other points. Point out that a "good" line of best fit should go through as many data points as possible. If it cannot go through all points, it should leave an approximately equal number of data points above and below the line.

TEACHING TIP You might want to discuss whether all sets of data will show a correlation. For instance, would it make sense to make a scatter plot showing students' grades vs. their height? Why?

LESSON 4.2

TEACHING TIP There will be times when the scales for the x- and y-axes will not be the same. Students should be able to choose an appropriate scale for each axis based on their table of values. You can use Example 4 on page 212 to discuss what increments should be used to create a reasonably-sized graph.

TEACHING TIP It makes sense to use negative and positive values of the independent variable for exercises such as Example 2 on page 211. However, real-life situations such as the one described in Example 4 might require a discussion as to what type of values are acceptable for the variables. For instance, if we were graphing students' age vs. height, would it make sense to use negative values?

LESSON 4.3

COMMON ERROR After completing Example 1 on page 218, some students might jump to the conclusion that when the equation of a line is given as $Ax + By = C$, its x- and y-intercepts will be B and A respectively. Show them an exercise such as Example 2 on page 219 where this is not true.

LESSON 4.4

COMMON ERROR When the graph of a line is given, students can find its slope by using a slope-triangle, eliminating the need for the slope formula. This is the way in which the slope formula was introduced on page 226. However, to use this method to find the slope it is absolutely necessary to adjust for any difference in the scales along the axes. The graph provided for Example 6 on page 229 can be used to show this fact.

TEACHING TIP Subscript notation will be new to many students and it can be confusing for them at first. Help your students by reading the variables in the slope formula such as x_1 as "x coordinate of the first point". You might want to write the slope formula labeling each of the variables, especially if you have students with special needs.

INCLUSION The idea of rate of change can be difficult to understand for some students. You can help them out by making a table of values for Example 6 on page 229. Take increments of 5 seconds to create your table and find the height above the ground for each value. You can now use your table to show your students that the change in height is the same every 5 seconds. The comparison between the change in height and the change in time is the rate of change.

LESSON 4.5

COMMON ERROR You might need to review the concepts of variable and constant for this Lesson. Many students believe that once the values of x and/or y are known, they will not change. (They think of variables as constants.) This misconception will prevent them from solving direct variation problems. Point out that when solving a certain problem, x and y can have an infinite number of different values, but the value of the constant of variation, k, stays the same.

Chapter Support

LESSON 4.5 (CONT.)

INCLUSION Help students with limited English proficiency to recognize and set up word problems involving direct variation. They should be able to tell if the problem involves direct variation just by looking for key words in the problem such as "...varies directly..." or "...direct variation model...". Practice with your students how to pick a variable for each quantity in the problem and how to translate sentences such as "...varies directly with..." into an equation writing a direct variation model.

LESSON 4.6

INCLUSION You might need to review the ideas of domain and range before completing Example 4 on page 243. Slowly explain what the restrictions in the domain mean, namely that each stage has a beginning point and an ending point in the graph. It is easy to graph each line by just finding those two points. Since the stages are consecutive, the ending point of stage 1 is the beginning point for stage 2. This fact can be used to graph faster. After completing the graph showing all stages of the flood, you can ask your students how many different line segments make up the graph.

LESSON 4.7

TEACHING TIP After learning to solve linear equations by graphing, students should be able to discuss the pros and cons of algebraic and graphing methods. Completing an example where the solution is not an integer number, such as $3x - 7 = 0$, can lead the discussion into the inaccuracy of using the graphing method. However, the graphing method can help students to make predictions more easily than by using algebraic techniques. If you have access to graphing calculators, you can demonstrate how they help to resolve the problem of inaccuracy inherent to the graphing method.

LESSON 4.8

COMMON ERROR Some students misunderstand the meaning of function notation. They think that $f(x)$ really means "f times x." If you ask these students to evaluate $f(x) = 3x + 1$ when $x = 2$, they might think something like "$2f = 7$, and therefore $f = 3.5$." Emphasize that $f(x)$ is just another way of writing the y-coordinate. You can help your students by teaching them to read an expression such as $f(2) = 7$ as "when the input of the function is 2, its output is 7." Point out that the only reason we use this new notation is to indicate that we are working with a function, so that we know that for each input there is just one output.

Outside Resources

BOOKS/PERIODICALS

VanDyke, Frances. "Relating to Graphs in Introductory Algebra." *Mathematics Teacher* (September 1994); pp. 427–432, 438, 439.

ACTIVITIES/MANIPULATIVES

Anderson, Edwin D. and Jim Nelson. "An Introduction to the Concept of Slope." *Mathematics Teacher* (January 1994); pp. 27–30, 37–41.

SOFTWARE

Dugdale, Sharon and David Kibbey. *Green Globs and Graphing Equations*. Introductory graphing concepts, tutorials, exploring graphs. Pleasantville, NY; Sunburst Communications.

VIDEOS

Algebra in Simplest Terms. Linear equations. Burlington, VT; Annenburg/CPB Collection, 1991.

Parent Guide for Student Success

For use with Chapter 4: Graphing Linear Equations and Functions

Chapter Overview One way that you can help your student succeed in Chapter 4 is by discussing the lesson goals in the chart below. When a lesson is completed, ask your student to interpret the lesson goals for you and to explain how the mathematics of the lesson relates to one of the key applications listed in the chart.

Lesson Title	Lesson Goals	Key Applications
4.1: Coordinates and Scatter Plots	Plot points in a coordinate plane and draw a scatter plot to make real-life predictions.	• Sports Equipment • Car Comparisons • American Flamingoes
4.2: Graphing Linear Equations	Graph the equation of a line using a table or a list of values. Graph horizontal and vertical lines.	• Internet Access • Landscaping Business • Triathlon
4.3: Quick Graphs Using Intercepts	Find the intercepts of the graph of a linear equation and use intercepts to make a quick graph of the equation.	• Zoo Fundraising • School Play • Movie Prices
4.4: The Slope of a Line	Find the slope of a line using two of its points and interpret slope as a rate of change.	• Parachuting • Road Grades • Cable Cars • Population Rates
4.5: Direct Variation	Write a linear equation that represents direct variation. Use a ratio to write an equation for direct variation.	• Zeppelins • Animal Studies • Violin Family
4.6: Quick Graphs Using Slope-Intercept Form	Graph and interpret a linear equation in slope-intercept form.	• City Planning • Snowstorms • Roller Coasters
4.7: Solving Linear Equations Using Graphs	Solve a linear equation graphically and use a graph to solve real-life problems.	• Nurses • Production Costs • Tourism
4.8: Functions and Relations	Identify when a relation is a function. Use function notation to represent real-life situations.	• Butterfly Migration • Zydeco Music • Masters Tournament

Study Strategy

Getting Your Questions Answered is the study strategy featured in Chapter 4 (see page 202). Encourage your student to make a list of questions whenever he or she doesn't understand something. If you can't answer the questions, have your student ask the teacher and then record the answers in his or her notebook.

NAME _____ DATE _____

Parent Guide for Student Success

For use with Chapter 4

Key Ideas Your student can demonstrate understanding of key concepts by working through the following exercises with you.

Lesson	Exercise
4.1	In preparation for her trip to Spain, Susan is practicing converting euros into United States dollars. The rate of currency exchange is 1.053 dollars per euro. Use a scatter plot to check these results she recorded: 25 euros to $26, 95 euros to $100, 120 euros to $140, and 210 euros to $221.
4.2	Find three different ordered pairs that are solutions of the equation $2x - 5y = 10$.
4.3	Find the x- and y-intercepts of the equation $7x - 4y = 28$.
4.4	Find the slope of the line passing through the points $(8, 1)$ and $(-2, -4)$.
4.5	The amount of gasoline used by a car varies directly with the distance traveled. A car uses 4 gallons of gasoline to travel 92 miles. How much gasoline is needed to travel 207 miles?
4.6	Decide whether the graphs of $2x - 3y = 9$ and $y = \frac{2}{3}x + 4$ are parallel. Explain your answer.
4.7	Write the equation $4 - 3x = 10$ in the form $ax + b = 0$. Write the related function $y = ax + b$. Then use a graph to solve the equation.
4.8	Evaluate $f(x) = 7x - 4$ when $x = 2$.

Home Involvement Activity

You will need: a yardstick, tape, a tennis ball, a piece of uncooked spaghetti

Directions: Tape the yardstick to the refrigerator or to a wall so that zero is at ground level. Pick a height such as 30 inches. Hold the tennis ball so that the top is at this height above ground. Drop the ball while your student records how high the top of the ball reaches during the bounce. Repeat two more times. Find the average height of the three bounces and plot the point defined by the drop height and the average bounce height. Repeat for at least seven other drop heights less than one yard. Then do the following:

• Describe the relationship between the bounce height and the drop height.

• The data points should fall in approximately a straight line. Use a piece of spaghetti to model that line. Estimate the slope and the y-intercept of the "line."

• Use the "spaghetti line" to predict how high the ball would bounce if dropped from the top of the yardstick. Then test your prediction by carrying out this drop.

Answers

4.1: 120 euros converts to $126; others are correct **4.2:** Answers may vary. *Sample answer:* $(0, -2)$, $(5, 0)$, $(-5, -4)$ **4.3:** $(4, 0)$ and $(0, -7)$ **4.4:** $\frac{1}{2}$ **4.5:** 9 gallons **4.6:** parallel; both have a slope of $\frac{2}{3}$ **4.7:** $3x + 6 = 0$; $y = 3x + 6$; -2 **4.8:** 10

Prerequisite Skills Review

For use before Chapter 4

EXAMPLE 1 *Writing Percents as Fractions and Decimals*

Write each percent as a fraction in lowest terms and as a decimal.

a. 34%   **b.** 5%

SOLUTION

% means "divided by 100":   Move the decimal point **left** two places:

a. $34\% = \dfrac{34}{100} = \dfrac{17}{50}$ and $34\% = 34\% = 0.34$

b. $5\% = \dfrac{5}{100} = \dfrac{1}{20}$ and $5\% = 05\% = 0.05$

Exercises for Example 1

Write each percent as a fraction in lowest terms and as a decimal.

1. 25%   **2.** 60%   **3.** 8%   **4.** 114%

EXAMPLE 2 *Working with Functions*

Use the function $y = \frac{1}{2}x + 10$, where $-10 \le x \le 10$.

a. Make an input-output table for the function.

b. Draw a line graph to represent the data in the input-output table.

c. Describe the domain and range of the function.

SOLUTION

a. You can choose any input values from -10 to 10 to make the input-output table. You should at least choose the input values -10 and 10.

Input	Function	Output
$x = -10$	$y = \frac{1}{2}(-10) + 10$	$y = 5$
$x = -5$	$y = \frac{1}{2}(-5) + 10$	$y = 7\frac{1}{2}$
$x = 0$	$y = \frac{1}{2}(0) + 10$	$y = 10$
$x = 5$	$y = \frac{1}{2}(5) + 10$	$y = 12\frac{1}{2}$
$x = 10$	$y = \frac{1}{2}(10) + 10$	$y = 15$

Once you have evaluated each input value, you can use a table to summarize your results.

Input, x	-10	-5	0	5	10
Output, y	5	$7\frac{1}{2}$	10	$12\frac{1}{2}$	15

Prerequisite Skills Review

For use before Chapter 4

b. To draw a line graph, write the values listed in the input-output table as ordered pairs. Then plot each ordered pair on the coordinate plane.

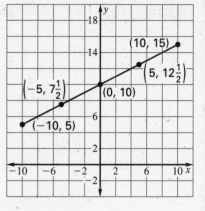

Ordered pairs

$(-10, 5)$

$\left(-5, 7\frac{1}{2}\right)$

$(0, 10)$

$\left(5, 12\frac{1}{2}\right)$

$(10, 15)$

c. Your table and graph show you that as x goes from -10 to 10 (the domain), y goes from 5 to 15 (the range).

Domain: $-10 \le x \le 10$ Range: $5 \le y \le 15$

Exercises for Example 2

For each function, make an input-output table, draw a line graph, and find the domain and range.

5. $y = -2x + 7$, where $x \ge 0$ **6.** $y = 3x - 5$, where $x \le 5$

EXAMPLE 3 *Evaluating Expressions*

Evaluate $\dfrac{3x - 2y}{y}$ when $x = 2$ and $y = -3$.

SOLUTION

$$\frac{3x - 2y}{y} = \frac{3(2) - 2(-3)}{-3} \qquad \leftarrow \text{Substitute the value of each variable.}$$

$$= \frac{6 - (-6)}{-3} \qquad \leftarrow \text{Simplify.}$$

$$= \frac{12}{-3}$$

$$= -4 \qquad \leftarrow \text{So, the value of the expression is } -4.$$

Exercises for Example 3

Evaluate each expression for the given values of the variables.

7. $\dfrac{x + y}{2x}$ when $x = 4$ and $y = -2$ **8.** $\dfrac{3xy}{y}$ when $x = -2$ and $y = -1$

NAME _____ DATE _____

Strategies for Reading Mathematics

For use with Chapter 4

Strategy: Reading Tables and Graphs in Algebra

Suppose a car averages about 30 miles per gallon of gasoline. How could you represent the function that shows how distance in miles (output) depends on gallons of gasoline (input)?

In Chapter 1 you learned that a function can be represented in more than one way. Your textbook frequently uses words, tables, and graphs together to help you gain a better understanding of algebra. Both the table and the graph below show that for every gallon of gasoline, the car travels about 30 miles.

Table

Input	Output
Gasoline in gallons, g	Distance in miles, d
0	0
1	30
2	60
3	90
4	120

Graph

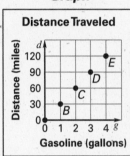

STUDY TIP

Reading a Table

When you are reading a table, use the headings to remind yourself what each number in the table represents. The table shows that a car traveled about 120 miles on 4 gallons of gasoline.

STUDY TIP

Reading a Graph

As you read a graph, be sure to read the title, headings, and graph labels to remind yourself what each point represents.

Questions

1. Which labeled point on the graph represents a distance of 60 miles? Which labeled point represents 3 gallons of gasoline? What does the labeled point E represent?

2. The labels of a table or graph can help you decide whether you can extend them to include other points. Think: Can a car have less than 0 gallons of gasoline? Can a car have $3\frac{1}{2}$ gallons of gasoline? Can a car with enough gasoline travel 185.2 miles? Can the table and graph be extended?

3. Do you think you can always read the values for every point on a graph exactly? Why or why not?

4. You can also use an equation to represent a function. Which equation describes this relationship between distance traveled and gasoline used: $d = 30g$, where $g \geq 0$, or $g = 30d$ where $d \geq 0$? Explain your reasoning. (*Hint:* What does each variable mean?)

NAME _____ DATE _____

Strategies for Reading Mathematics

For use with Chapter 4

Chapter Support

Visual Glossary

The Study Guide on page 202 lists the key vocabulary for Chapter 4 as well as review vocabulary from previous chapters. Use the page references on page 202 or the Glossary in the textbook to review key terms from prior chapters. Use the visual glossary below to help you understand some of the key vocabulary in Chapter 4. You may want to copy these diagrams into your notebook and refer to them as you complete the chapter.

GLOSSARY

coordinate plane (p. 203) A plane formed by two real number lines (called *axes*) that intersect at a right angle. The intersection is point $(0, 0)$, or the *origin*.

ordered pair (p. 203) A pair of numbers used to identify a point in a coordinate plane.

graph of a linear equation (p. 210) The set of all points (x, y) that are solutions of the linear equation.

y-intercept (p. 218) The y-coordinate of a point where a graph crosses the y-axis.

x-intercept (p. 218) The x-coordinate of a point where a graph crosses the x-axis.

slope (p. 226) The number of units a nonvertical line rises or falls for each unit of horizontal change from left to right.

Graphing in a Coordinate Plane

The idea that an ordered pair of numbers, such as $(4, -3)$, can be pictured by a point in a plane is very valuable in algebra.

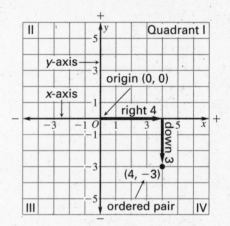

Graphing Linear Equations

The graph of an equation can show you whether a relationship is linear or nonlinear. The equation $y = 2x + 3$ is a linear equation because its graph is a straight line.

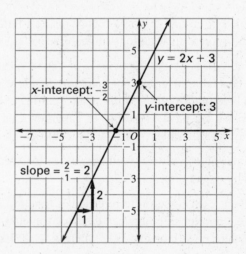

TEACHER'S NAME _____ CLASS _____ ROOM _____ DATE _____

Lesson Plan

1-day lesson (See *Pacing the Chapter,* TE pages 200C–200D)   For use with pages 203–209

 GOALS 1. **Plot points in a coordinate plane.**
2. **Draw a scatter plot and make predictions about real-life situations.**

State/Local Objectives _____

_____

✓ **Check the items you wish to use for this lesson.**

STARTING OPTIONS
_____ Prerequisite Skills Review: CRB pages 5–6
_____ Strategies for Reading Mathematics: CRB pages 7–8
_____ Warm-Up or Daily Homework Quiz: TE pages 203 and 185, CRB page 11, or Transparencies

TEACHING OPTIONS
_____ Motivating the Lesson: TE page 204
_____ Lesson Opener (Application): CRB page 12 or Transparencies
_____ Graphing Calculator Activity with Keystrokes: CRB pages 13–15
_____ Examples 1–3: SE pages 203–205
_____ Extra Examples: TE pages 204–205 or Transparencies
_____ Technology Activity: SE page 209
_____ Closure Question: TE page 205
_____ Guided Practice Exercises: SE page 206

APPLY/HOMEWORK
Homework Assignment
_____ Basic   10–26 even, 27–32, 35–38, 42–52 even
_____ Average   10–26 even, 27–32, 35–38, 42–52 even
_____ Advanced   10–26 even, 27–32, 35–38, 42–52 even

Reteaching the Lesson
_____ Practice Masters: CRB pages 16–18 (Level A, Level B, Level C)
_____ Reteaching with Practice: CRB pages 19–20 or Practice Workbook with Examples
_____ Personal Student Tutor

Extending the Lesson
_____ Applications (Interdisciplinary): CRB page 22
_____ Challenge: SE page 208; CRB page 23 or Internet

ASSESSMENT OPTIONS
_____ Checkpoint Exercises: TE pages 204–205 or Transparencies
_____ Daily Homework Quiz (4.1): TE page 208, CRB page 26, or Transparencies
_____ Standardized Test Practice: SE page 208; TE page 208; STP Workbook; Transparencies

Notes _____

_____

_____

TEACHER'S NAME _____ CLASS _____ ROOM _____ DATE _____

Lesson Plan for Block Scheduling

Half-day lesson (See *Pacing the Chapter*, TE pages 200C–200D)   For use with pages 203–209

GOALS
1. **Plot points in a coordinate plane.**
2. **Draw a scatter plot and make predictions about real-life situations.**

State/Local Objectives _____

_____

_____

✓ **Check the items you wish to use for this lesson.**

STARTING OPTIONS
____ Prerequisite Skills Review: CRB pages 5–6
____ Strategies for Reading Mathematics: CRB pages 7–8
____ Warm-Up or Daily Homework Quiz: TE pages 203 and
          185, CRB page 11, or Transparencies

TEACHING OPTIONS
____ Motivating the Lesson: TE page 204
____ Lesson Opener (Application): CRB page 12 or Transparencies
____ Graphing Calculator Activity with Keystrokes: CRB pages 13–15
____ Examples 1–3: SE pages 203–205
____ Extra Examples: TE pages 204–205 or Transparencies
____ Technology Activity: SE page 209
____ Closure Question: TE page 205
____ Guided Practice Exercises: SE page 206

APPLY/HOMEWORK
Homework Assignment
____ Block Schedule: 10–26 even, 27–32, 35–38, 42–52 even

Reteaching the Lesson
____ Practice Masters: CRB pages 16–18 (Level A, Level B, Level C)
____ Reteaching with Practice: CRB pages 19–20 or Practice Workbook with Examples
____ Personal Student Tutor

Extending the Lesson
____ Applications (Interdisciplinary): CRB page 22
____ Challenge: SE page 208; CRB page 23 or Internet

ASSESSMENT OPTIONS
____ Checkpoint Exercises: TE pages 204–205 or Transparencies
____ Daily Homework Quiz (4.1): TE page 208, CRB page 26, or Transparencies
____ Standardized Test Practice: SE page 208; TE page 208; STP Workbook; Transparencies

Notes _____

_____

_____

CHAPTER PACING GUIDE	
Day	**Lesson**
1	Assess Ch. 3; **4.1 (all)**
2	4.2 (all)
3	4.3 (all)
4	4.4 (all)
5	4.5 (all); 4.6 (begin)
6	4.6 (end); 4.7 (all)
7	4.8 (all)
8	Review/Assess Ch. 4

Lesson 4.1

NAME _____ DATE _____

WARM-UP EXERCISES

For use before Lesson 4.1, pages 203–209

Use the formula $F = \frac{9}{5}C + 32$ to find the following temperatures.

1. What Celsius temperature is equivalent to 32°F?

2. What Fahrenheit temperature is equivalent to 225°C?

3. What Celsius temperature is equivalent to −40°F?

DAILY HOMEWORK QUIZ

For use after Lesson 3.8, pages 180–186

1. What is the average speed of a plane that flies 2629 mi in 5.5 h?

2. A moving company one year had an income of $930,500. The number of moving jobs was 882. What was the average rate per job?

3. Find the percent. Round to the nearest whole percent. Tip of $5.50 on a restaurant bill of $32.50

4. A survey of telephone customers in one city asked, "If you could save $15 per month on long-distance calls, would you change your long-distance company?" Of those surveyed, 76 answered yes. About how many customers took part in the survey?

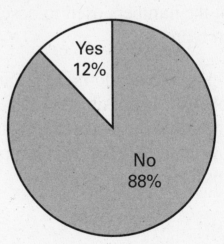

NAME _____ DATE _____

Application Lesson Opener

For use with pages 203–208

You decide to start babysitting to earn some money for a trip you are taking this summer. The graph shows the amount of money you can earn.

1. What do the numbers on the *x*-axis represent?

2. What do the numbers on the *y*-axis represent?

3. To move to each point on the graph, you start at the origin, which is labeled *O*. Find this point on the graph.

4. From the origin, move right or left along the *x*-axis, and then move up or down. Find point *A* on the graph. Describe how to get to this point from the origin. What do the numbers used to describe point *A* represent?

5. Describe how to get to the other points on the graph from the origin. Explain what the numbers used to describe each point represent.

6. Is it possible to add more points to the graph? If it is possible, add at least two points. How did you decide which points to add? What do these points represent?

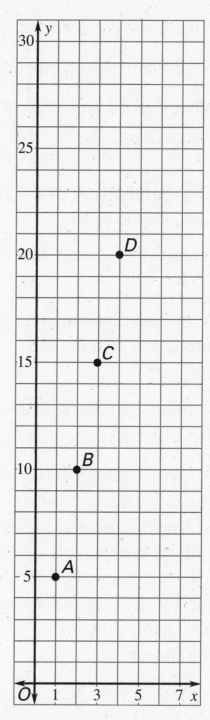

Algebra 1
Chapter 4 Resource Book

NAME _____ DATE _____

Graphing Calculator Activity

For use with pages 203–209

GOAL To tell which quadrant a point is in by looking at the signs of its coordinates.

By observing the positive and negative signs of the *x*- and *y*-coordinates in an ordered pair, it is possible to predict which one of the four *quadrants* a point will be located in.

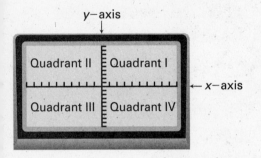

y–axis

← x–axis

Activity

1 Observe the graph at the left. It is divided into four quadrants by the *x*- and *y*-axes.

2 Listed below are a series of *ordered pairs*. The first number is the *x*-value and the second number is the *y*-value.

$(-3, 2), (9, 3), (-5, -1), (7, -6), (5, -5), (-2, 4)$
$(-1, -8), (3, 8), (-9, -9), (6, -7), (4, 4), (-2, 5)$

Predict which quadrant each point will be located in. Write your prediction on a piece of paper.

3 Use your graphing calculator to enter the *x*-values in List 1 and the *y*-values in List 2.

4 Plot the points in a coordinate plane like the one shown.

5 Use the TRACE feature to locate each point and check your prediction.

Exercises

1. Write a sentence to describe each of the following.

  a. The signs of the *x*-value and the *y*-value of a point located in Quadrant I

  b. The signs of the *x*-value and the *y*-value of a point located in Quadrant II

  c. The signs of the *x*-value and the *y*-value of a point located in Quadrant III

  d. The signs of the *x*-value and the *y*-value of a point located in Quadrant IV

2. Does the plot of an ordered pair of real numbers always lie in one of the four quadrants? Plot the following points to help you make a conclusion.

  a. $(0, 3)$ **b.** $(2, 0)$ **c.** $(-4, 0)$ **d.** $(0, -6)$

See page 14 for keystrokes.

NAME _____ DATE _____

Graphing Calculator Activity

For use with pages 203–209

TI-82

STAT 1

Enter *x*-values in L1.

(-) 3 ENTER 9 ENTER (-) 5 ENTER 7
ENTER 5 ENTER (-) 2 ENTER (-) 1
ENTER 3 ENTER (-) 9 ENTER 6 ENTER
4 ENTER (-) 2 ENTER

Enter *y*-values in L2.

2 ENTER 3 ENTER (-) 1 ENTER (-) 6
ENTER (-) 5 ENTER 4 ENTER (-) 8
ENTER 8 ENTER (-) 9 ENTER (-) 7
ENTER 4 ENTER 5 ENTER

2nd [STAT PLOT] 1

Choose the following.

On; Type: ⋰ ; Xlist: L1; Ylist: L2; Mark: ▫

ZOOM 6

TI-83

STAT 1

Enter *x*-values in L1.

(-) 3 ENTER 9 ENTER (-) 5 ENTER 7
ENTER 5 ENTER (-) 2 ENTER (-) 1
ENTER 3 ENTER (-) 9 ENTER 6 ENTER
4 ENTER (-) 2 ENTER

Enter *y*-values in L2.

2 ENTER 3 ENTER (-) 1 ENTER (-) 6
ENTER (-) 5 ENTER 4 ENTER (-) 8
ENTER 8 ENTER (-) 9 ENTER (-) 7
ENTER 4 ENTER 5 ENTER

2nd [STAT PLOT] 1

Choose the following.

On; Type: ⋰ ; Xlist: L1; Ylist: L2; Mark: ▫

ZOOM 6

SHARP EL-9600c

STAT [A]

Enter *x*-values in L1.

(-) 3 ENTER 9 ENTER (-) 5 ENTER 7
ENTER 5 ENTER (-) 2 ENTER (-) 1
ENTER 3 ENTER (-) 9 ENTER 6 ENTER
4 ENTER (-) 2 ENTER

Enter *y*-values in L2.

2 ENTER 3 ENTER (-) 1 ENTER (-) 6
ENTER (-) 5 ENTER 4 ENTER (-) 8
ENTER 8 ENTER (-) 9 ENTER (-) 7
ENTER 4 ENTER 5 ENTER

2ndF [STAT PLOT] [A]

Choose the following.

on; DATA XY; List X: L1; List Y: L2

2ndF [STAT PLOT] [G] 3 ZOOM [A]5

CASIO CFX-9850GA PLUS

From the main menu, choose STAT.
Enter *x*-values in List 1.

(-) 3 EXE 9 EXE (-) 5 EXE 7 EXE 5
EXE (-) 2 EXE (-) 1 EXE 3 EXE (-) 9
EXE 6 EXE 4 EXE (-) 2 EXE

Enter *y*-values in List 2.

2 EXE 3 EXE (-) 1 EXE (-) 6 EXE (-) 5
EXE 4 EXE (-) 8 EXE 8 EXE (-) 9 EXE
(-) 7 EXE 4 EXE 5
SHIFT F3 F3 EXIT SHIFT [SET UP] F2
EXIT F1 F6

Choose the following.
Graph Type: Scatter; XList: List1; YList: List 2;
Frequency: 1; Mark Type: ▫
EXIT F1

Graphing Calculator Activity Keystrokes

For use with Technology Activity 4.1 on page 209

TI-82

STAT 1

Enter *x*-values in L1.

10 ENTER 11 ENTER 12 ENTER

13 ENTER 14 ENTER 15 ENTER 16 ENTER

17 ENTER

Enter *y*-values in L2.

7.95 ENTER 7.53 ENTER 7.18 ENTER

6.83 ENTER 6.43 ENTER 6.33 ENTER

6.13 ENTER 6.10 ENTER

WINDOW 0 ENTER 20 ENTER 5 ENTER

0 ENTER 10 ENTER 2 ENTER

2nd [STAT PLOT] 1

Choose the following.

On; Type: ⠾ ; Xlist: L1; Ylist: L2; Mark: ▫

GRAPH

TI-83

STAT 1

Enter *x*-values in L1.

10 ENTER 11 ENTER 12 ENTER

13 ENTER 14 ENTER 15 ENTER 16 ENTER

17 ENTER

Enter *y*-values in L2.

7.95 ENTER 7.53 ENTER 7.18 ENTER

6.83 ENTER 6.43 ENTER 6.33 ENTER

6.13 ENTER 6.10 ENTER

WINDOW 0 ENTER 20 ENTER 5 ENTER

0 ENTER 10 ENTER 2 ENTER

2nd [STAT PLOT] 1

Choose the following.

On; Type: ⠾ ; Xlist: L1; Ylist: L2; Mark: ▫

GRAPH

SHARP EL-9600c

STAT [A]

Enter *x*-values in L1.

10 ENTER 11 ENTER 12 ENTER

13 ENTER 14 ENTER 15 ENTER 16 ENTER

17 ENTER

Enter *y*-values in L2.

7.95 ENTER 7.53 ENTER 7.18 ENTER

6.83 ENTER 6.43 ENTER 6.33 ENTER

6.13 ENTER 6.10 ENTER

WINDOW 0 ENTER 20 ENTER 5 ENTER

0 ENTER 10 ENTER 2 ENTER

2ndF [STAT PLOT] [A] ENTER

Choose the following.

on; DATA XY; List X: L1; List Y: L2

2ndF [STAT PLOT] [G] 3 GRAPH

CASIO CFX-9850GA PLUS

From the main menu, choose STAT.
Enter *x*-values in List 1.

10 EXE 11 EXE 12 EXE 13 EXE 14 EXE

15 EXE 16 EXE 17 EXE

Enter *y*-values in List 2.

7.95 EXE 7.53 EXE 7.18 EXE 6.83 EXE

6.43 EXE 6.33 EXE 6.13 EXE 6.10 EXE

SHIFT F3 0 EXE 20 EXE 5 EXE 0 EXE

10 EXE 2 EXE EXIT SHIFT [SET UP] F2

EXIT F1 F6

Choose the following.

Graph Type: Scatter; XList: List1; YList: List 2;
Frequency: 1; Mark Type: ▫

EXIT F1

NAME _____ DATE _____

Practice A

For use with pages 203–208

Write the ordered pairs that correspond to the points labeled *A*, *B*, *C*, and *D* in the coordinate plane.

1.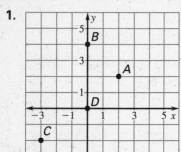

2.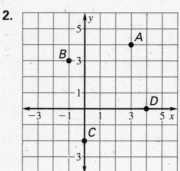

3.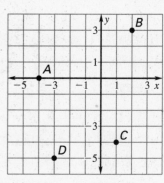

Plot and label the ordered pairs in a coordinate plane.

4. $(2, 2), (2, 4), (2, 5)$ **5.** $(3, 2), (2, 1), (4, 0)$ **6.** $(-3, 1), (-4, 1), (2, -1)$

7. $(-5, -2), (-5, 0), (-3, 2)$ **8.** $(0, 2), (3, -3), (-1, -3)$ **9.** $(-1, 1), (0, -2), (3, 4)$

Without plotting the point, tell whether it is in Quadrant I, Quadrant II, Quadrant III, or Quadrant IV.

10. $(3, 4)$ **11.** $(5, -2)$ **12.** $(2, -5)$ **13.** $(-1, -3)$

14. $(-4, 3)$ **15.** $(-2, -2)$ **16.** $(6, 1)$ **17.** $(-2, 4)$

18. *Hourly Pay*   The table shows the number of hours worked and the corresponding pay in dollars. Make a scatter plot of the data. Let each ordered pair have the form (h, d).

h	1	2	3	5	8
d	4.50	9.00	13.50	22.50	36.00

19. *Yards to Feet*   The table shows some measurements in yards and the corresponding measurement in feet. Make a scatter plot of the data. Let each ordered pair have the form (y, f).

y	1	5	10	15	20
f	3	15	30	45	60

20. *Basketball*   The following table shows the heights (in inches) of players on a high school basketball team and how many players are each height. Make a scatter plot of the data. Use the horizontal axis to represent the height.

Height (in inches)	69	70	71	72	73	74	75	76	77
Number of players	1	0	2	5	3	2	0	0	1

NAME _____ DATE _____

Practice B

For use with pages 203–208

**Write the ordered pairs that correspond to the points labeled *A, B, C,*
and *D* in the coordinate plane.**

1.
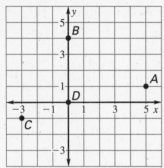

2.
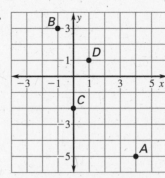

3.
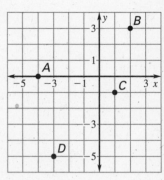

Plot and label the ordered pairs in a coordinate plane.

4. $(3, 6), (-2, 5), (2, 2)$

5. $(-3, -3), (2, -5), (1, 0)$

6. $(3, -1), (-4, -1), (-1, 6)$

7. $(5, 2), (-5, 0), (-3, 2)$

8. $(0, 3), (4, -4), (-2, -4)$

9. $(-1, 1), (0, -1), (4, 2)$

10. *Inches to Centimeters* The table shows some measurements in inches and
the corresponding measurement in centimeters. Make a scatter plot of the
data. Let each ordered pair have the form (i, c).

i	1	5	10	15	20
c	2.54	12.7	25.4	38.1	50.8

11. *Quiz Grades* The following table shows various quiz grades for an algebra
class and how many students received each grade. Make a scatter plot of the
data. Use the horizontal axis to represent the grade.

Grade	0	1	2	3	4	5	6	7	8	9	10
Number of students	0	0	2	1	1	2	4	3	12	7	4

12. *U.S. Representatives* The 1995 population, *P* (in millions), for eight states
is shown in the table below. The number of U.S. representatives, *R*, for each
state is given. Make a scatter plot of the data.

State	ID	NE	AR	CT	MD	WA	MA	VA
Population, P (in millions)	1.2	1.6	2.5	3.3	5.0	5.4	6.1	6.6
Number of U.S. representatives, R	2	3	4	6	8	9	10	11

13. *Interpreting Data* In the Population vs. U.S. representatives graph you
made in Exercise 12, describe the relationship between population and the
number of U.S. representatives.

Practice C

For use with pages 203–208

Write the ordered pairs that correspond to the points labeled *A, B, C,*
and *D* **in the coordinate plane.**

1. 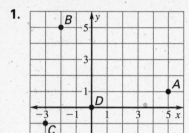   **2.** 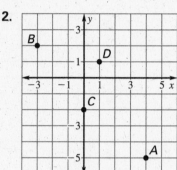   **3.**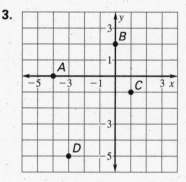

Plot and label the ordered pairs in a coordinate plane.

4. $(1, 4), (0, 2), (2, -5)$

5. $(-5, 1), (-2, -4), (-1, 0)$

6. $(0, -5), (-3, 2), (6, 0)$

7. $(-3, -6), (-3, -3), (3, -3)$

8. $(-2, 4), (-4, -1), (6, -3)$

9. $(-4, 0), (-1, 5), (3, 3)$

10. *Movie Rental*   The following table shows the amount spent, *S* (in billions of
dollars), on movie rentals for each year from 1990 through 1995, where
$t = 0$ represents 1990. Make a scatter plot of the data. Let each ordered pair
have the form (t, S).

t	0	1	2	3	4	5
S	5.0	5.0	5.5	6.0	6.6	6.7

11. *Profit*   The table shows the profit, *P* (in thousands of dollars), of a small sport-
ing goods store for each month, where $t = 1$ represents the month of January.
Make a scatter plot of the data. Describe the profit earned over the year.

Month, t	1	2	3	4	5	6	7	8	9	10	11	12
Profit, P (in $1000s)	3	2	2	3	4	7	8	7	5	3	3	7

12. *U.S. Representatives*   The 1996 population, *P* (in millions), for seven states
is shown in the table below. The number of U.S. representatives, *R*, for each
state is given. Make a scatter plot of the data.

State	AK	OR	MN	NC	MI	IL	FL
Population, P (in millions)	0.6	3.2	4.7	7.3	9.6	11.8	14.4
Number of U.S. representatives, R	1	5	8	12	16	20	23

13. *Interpreting Data*   In the Population vs. U.S. representatives graph you
made in Exercise 12, describe the relationship between population and the
number of U.S. representatives.

14. *Challenge*   Predict the number of U.S. representatives for a state with a
population of 18.2 million. Explain your reasoning.

NAME _____ DATE _____

Reteaching with Practice

For use with pages 203–208

GOAL Plot points in a coordinate plane, draw a scatter plot, and make predictions about real-life situations.

VOCABULARY

A **coordinate plane** is formed by two real number lines that intersect at a right angle.

Each point in a coordinate plane corresponds to an **ordered pair** of real numbers. The first number is the **x-coordinate** and the second number is the **y-coordinate.**

A **scatter plot** is a graph containing several points that represent real-life data.

EXAMPLE 1 *Plotting Points in a Coordinate Plane*

Plot and label the following ordered pairs in a coordinate plane.

a. $(3, -2)$ **b.** $(-4, 3)$

SOLUTION

To plot a point, you move along the horizontal and vertical lines in the coordinate plane and mark the location that corresponds to the ordered pair.

a. To plot the point $(3, -2)$, start at the origin. Move 3 units to the right and 2 units down.

b. To plot the point $(-4, 3)$, start at the origin. Move 4 units to the left and 3 units up.

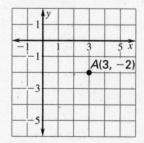 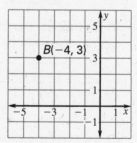

Exercises for Example 1

Plot and label the ordered pairs in a coordinate plane.

1. $A(5, 4), B(-3, 0), C(-1, -2)$ **2.** $A(-3, 2), B(0, 0), C(2, -2)$

3. $A(0, -4), B(3, 5), C(3, -1)$ **4.** $A(-1, -2), B(5, -2), C(-4, 0)$

5. $A(-1, 3), B(2, 0), C(3, -2)$ **6.** $A(2, 4), B(-2, 5), C(0, 3)$

NAME _____ DATE _____

Reteaching with Practice

For use with pages 203–208

EXAMPLE 2 *Sketching a Scatter Plot*

The table below gives the U.S. postal rates (in cents) for first-class mail, based on the weight (in ounces) of the mail. Draw a scatter plot of the data and predict the postal rate for a piece of mail that weighs 8 ounces.

Weight (ounces)	1	2	3	4	5
Rate (cents)	33	55	77	99	121

SOLUTION

❶ Rewrite the data in the table as a list of ordered pairs.

$(1, 33), (2, 55), (3, 77), (4, 99), (5, 121)$

❷ Draw a coordinate plane. Put weight w on the horizontal axis and rate r on the vertical axis.

❸ Plot the points.

❹ From the scatter plot, you can see that the points follow a pattern. By extending the pattern, you can predict that the postal rate for an 8 ounce piece of mail is about 187 cents, or $1.87.

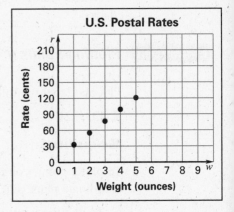

Exercises for Example 2

In Exercises 7 and 8, make a scatter plot of the data. Use the horizontal axis to represent time.

7.

Year	1997	1998	1999	2000
Members	74	81	89	95

8.

Month	Jan.	Apr.	Aug.	Dec.
Adults	22	30	15	42

In Exercises 9 and 10, use a scatter plot to see if the given information is correct. If not, explain how the data should be changed. Use the horizontal axis to represent quarts in Exercise 9 and hours in Exercise 10.

9.

Quarts	3.0	4.0	5.0	6.0
Gallons	0.75	1.0	1.3	1.5

10.

Hours	3	5	6	8
Rental charge (dollars)	14	20	24	32

NAME _____ DATE _____

Quick Catch-Up for Absent Students

For use with pages 203–209

The items checked below were covered in class on (date missed) _____

Lesson 4.1: Coordinates and Scatter Plots

____ **Goal 1:** Plot points in a coordinate plane. (p. 203)

Material Covered:

____ Example 1: Plotting Points in a Coordinate Plane

Vocabulary:

coordinate plane, p. 203   ordered pair, p. 203
x-coordinate, p. 203   *y*-coordinate, p. 203
x-axis, p. 203   *y*-axis, p. 203
origin, p. 203   quadrants, p. 203

____ **Goal 2:** Draw a scatter plot and make predictions about real-life situations. (pp. 204–205)

Material Covered:

____ Example 2: Making a Scatter Plot

____ Example 3: Making Predictions from a Scatter Plot

Vocabulary:

scatter plot, p. 204

Activity 4.1: Graphing a Scatter Plot (p. 209)

____ **Goal:** Make a scatter plot using a graphing calculator or a computer.

____ Student Help: Keystroke Help

____ Other (specify) _____

_____

Homework and Additional Learning Support

____ Textbook (specify) pp. 206–208 _____

_____

____ *Reteaching with Practice* worksheet (specify exercises)_____

____ *Personal Student Tutor* for Lesson 4.1

NAME _____ DATE _____

Interdisciplinary Application

For use with pages 203–208

Mammals

BIOLOGY Cows, dogs, humans, and lions are all mammals. Mammals are different from most other types of animals in five ways.

- Mammals have hair at some time in their lives.
- Mammals are warm-blooded. This means that the body temperatures of mammals are about the same all of the time, even though the temperature of their environment changes.
- Mammals have brains that are larger and better developed than other animals.
- Mammals train and protect their young more than other animals.
- Mammals nurse their babies.

Before a mammal can nurse its baby, the mother carries its unborn young while it develops from conception to birth. This is called the gestation period. The length of the gestation period differs with the species, and may even vary with individual births of the same animal. The following table shows a mammal, its average gestation period (in days), and the average birth weight (in pounds).

Mammal	Cow	Dog	Elephant	Giraffe	Horse	Human	Lion	Mouse	Rabbit
Average gestation period (days)	284	61	641	410	337	267	108	19	31
Average birth weight (pounds)	50	0.5	243	132	50	7.5	3.5	0.0025	0.125

1. Make a scatter plot of the average gestation periods and the average birth weights for the nine mammals. Use the horizontal axis to represent the gestation period.

2. What is the heaviest average birth weight shown on the scatter plot? What is the lightest?

3. Describe the relationship between the average gestation period and the average birth weight.

Algebra 1
Chapter 4 Resource Book

NAME _____ DATE _____

Challenge: Skills and Applications

For use with pages 203–208

Make a scatter plot to investigate each relationship suggested by the data in the table. Use your scatter plots to answer the questions in Exercises 1–3.

State	A	B	C	D	E	F
Percent of 8th graders proficient in mathematics (1992)	64%	48%	78%	48%	57%	71%
Graduation rate (1995)	75%	57%	87%	64%	72%	82%
Expenditures per pupil (1995)	$8817	$5193	$4775	$4586	$5327	$6930

1. What is the relationship between a state's 8th-grade mathematics achievement in 1992 and its graduation rate in 1995?

2. What is the relationship between a state's expenditure per pupil in 1995 and its graduation rate in 1995?

3. The data in the table provide two indicators of a state's success in educating its precollege students and one indicator of money spent on education. What conclusions can you draw about success and money spent from the scatter plots you drew?

4. Use the scatter plot for Exercise 1 to make the following prediction. *If 50% of a state's 8th graders were proficient in mathematics in 1992, about what percent would you expect to have graduated in 1995?*

TEACHER'S NAME _____ CLASS _____ ROOM _____ DATE _____

Lesson Plan

2-day lesson (See *Pacing the Chapter*, TE pages 200C–200D)   **For use with pages 210–217**

GOALS
1. **Graph a linear equation using a table or a list of values.**
2. **Graph horizontal and vertical lines.**

State/Local Objectives _____

_____

✓ **Check the items you wish to use for this lesson.**

STARTING OPTIONS

____ Homework Check: TE page 206; Answer Transparencies

____ Warm-Up or Daily Homework Quiz: TE pages 210 and 208, CRB page 26, or Transparencies

TEACHING OPTIONS

____ Motivating the Lesson: TE page 211

____ Lesson Opener (Activity): CRB page 27 or Transparencies

____ Examples: Day 1: 1–3, SE pages 210–212; Day 2: 4–6, SE pages 212–213

____ Extra Examples: Day 1: TE pages 211–212 or Transp.; Day 2: TE pages 212–213 or Transp.;
                         Internet

____ Closure Question: TE page 213

____ Guided Practice: SE page 214; Day 1: Exs. 1, 2, 4–10; Day 2: Ex. 3, 11

APPLY/HOMEWORK

Homework Assignment

____ Basic   Day 1: 12–35, 36–50 even; Day 2: 52–55, 60–70, 74–78, 85, 90, 95

____ Average   Day 1: 12–35, 36–50 even; Day 2: 52–55, 60–70, 74–78, 85, 90, 95

____ Advanced   Day 1: 12–35, 36–50 even; Day 2: 52–56, 60–66, 71–80, 85, 90, 95

Reteaching the Lesson

____ Practice Masters: CRB pages 28–30 (Level A, Level B, Level C)

____ Reteaching with Practice: CRB pages 31–32 or Practice Workbook with Examples

____ Personal Student Tutor

Extending the Lesson

____ Applications (Real-Life): CRB page 34

____ Challenge: SE page 217; CRB page 35 or Internet

ASSESSMENT OPTIONS

____ Checkpoint Exercises: Day 1: TE page 211 or Transp.; Day 2: TE pages 212–213 or Transp.

____ Daily Homework Quiz (4.2): TE page 217, CRB page 38, or Transparencies

____ Standardized Test Practice: SE page 217; TE page 217; STP Workbook; Transparencies

Notes _____

_____

_____

Lesson Plan for Block Scheduling

1-day lesson (See *Pacing the Chapter*, TE pages 200C–200D)   **For use with pages 210–217**

GOALS 1. **Graph a linear equation using a table or a list of values.**
2. **Graph horizontal and vertical lines.**

State/Local Objectives _____

_____

_____

✓ **Check the items you wish to use for this lesson.**

STARTING OPTIONS

____ Homework Check: TE page 206; Answer Transparencies
____ Warm-Up or Daily Homework Quiz: TE pages 210 and
       208, CRB page 26, or Transparencies

TEACHING OPTIONS

____ Motivating the Lesson: TE page 211
____ Lesson Opener (Activity): CRB page 27 or Transparencies
____ Examples 1–6: SE pages 210–213
____ Extra Examples: TE pages 211–213 or Transparencies; Internet
____ Closure Question: TE page 213
____ Guided Practice Exercises: SE page 214

APPLY/HOMEWORK

Homework Assignment
____ Block Schedule: 12–35, 36–50 even, 52–55, 60–70, 74–78, 85, 90, 95

Reteaching the Lesson
____ Practice Masters: CRB pages 28–30 (Level A, Level B, Level C)
____ Reteaching with Practice: CRB pages 31–32 or Practice Workbook with Examples
____ Personal Student Tutor

Extending the Lesson
____ Applications (Real-Life): CRB page 34
____ Challenge: SE page 217; CRB page 35 or Internet

ASSESSMENT OPTIONS

____ Checkpoint Exercises: TE pages 211–213 or Transparencies
____ Daily Homework Quiz (4.2): TE page 217, CRB page 38, or Transparencies
____ Standardized Test Practice: SE page 217; TE page 217; STP Workbook; Transparencies

Notes _____

_____

_____

_____

CHAPTER PACING GUIDE	
Day	**Lesson**
1	Assess Ch. 3; 4.1 (all)
2	**4.2 (all)**
3	4.3 (all)
4	4.4 (all)
5	4.5 (all); 4.6 (begin)
6	4.6 (end); 4.7 (all)
7	4.8 (all)
8	Review/Assess Ch. 4

Lesson 4.2

Algebra 1
Chapter 4 Resource Book

Available as
a transparency

NAME _____ DATE _____

WARM-UP EXERCISES

For use before Lesson 4.2, pages 210–217

Rewrite each equation in function form by solving for *y*.

1. $2x + y = 10$ **2.** $6x - 3y = -3$

Find the value of *y* when *x* = −3.

3. $y = x - 7$ **4.** $y = -5x + 1$

···

DAILY HOMEWORK QUIZ

For use after Lesson 4.1, pages 203–209

1. Use the graph. Write the ordered pairs that correspond to points *A*, *C*, and *E*.

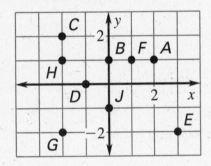

2. The heights *H* and weights *W* of 5 students are given in the table. Make a scatter plot of the data. Use the horizontal axis to represent heights.

H (in.)	63	75	74	66	70
W (lb)	112	176	186	131	138

NAME ——————————————— DATE ————

Activity Lesson Opener

For use with pages 210–217

SET UP: Work in a group.

1. Find the graph assigned to your group.

Graph 1: $2x - y = 2$

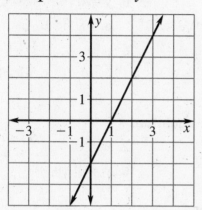

Graph 2: $x + y = 4$

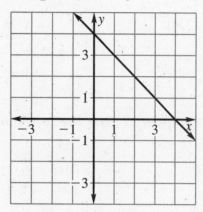

Graph 3: $x - 2y = 2$

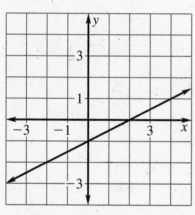

Graph 4: $x + 2y = -3$

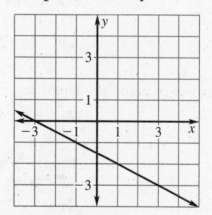

2. As your teacher calls out one of the ordered pairs at random, plot the corresponding point on your graph. If the point lies on your line, your group scores one point. The first group to score four points wins!

$(0, 4)$	$(3, -3)$	$(-1, -1)$	$(-2, -2)$	$(0, -2)$
$(0, -1)$	$(3, 4)$	$(1, 3)$	$(2, 2)$	$(1, 0)$
$(3, 1)$	$(2, 0)$	$(4, 1)$	$(-3, 0)$	$(1, -2)$

3. Substitute the coordinates of the points you have drawn on your graph into the given equation. What do you find?

Practice A

For use with pages 210–217

Decide which of the two points lies on the graph of the line.

1. $x + y = 8$
   a. $(2, 4)$ **b.** $(2, 6)$

2. $2x + y = 8$
   a. $(2, 2)$ **b.** $(3, 2)$

3. $y - x = 2$
   a. $(5, 3)$ **b.** $(3, 5)$

4. $x = 4$
   a. $(4, 2)$ **b.** $(2, 4)$

5. $y = -3$
   a. $(-3, 2)$ **b.** $(3, -3)$

6. $y = 0$
   a. $(0, 3)$ **b.** $(-1, 0)$

7. $y = x - 2$
   a. $(4, 6)$ **b.** $(6, 4)$

8. $y = x + 3$
   a. $(-2, 1)$ **b.** $(1, -2)$

9. $y = -3x + 1$
   a. $(0, 1)$ **b.** $(1, 4)$

Find three different ordered pairs that are solutions of the equation.

10. $y = x - 5$

11. $x = -2$

12. $y = 1$

13. $y = -x + 4$

14. $y = -3x - 4$

15. $y = 2(x + 4)$

Rewrite the equation in function form.

16. $-x + y = 6$

17. $x + y = -2$

18. $-x + y = -2$

19. $-2x + y = -4$

20. $3x - y = 1$

21. $-2x + y = 0$

22. $4x + 2y = 1$

23. $-9x + 3y = -6$

24. $-2x - 4y = 3$

Use a table of values to graph the equation.

25. $y = x + 3$

26. $y = x - 2$

27. $y = 2x + 3$

28. $y = 6$

29. $x = -1$

30. $x = 0$

31. $y = -x$

32. $y = \frac{2}{3}x + 6$

33. $y = \frac{1}{2}x + 4$

34. $y = 2 - x$

35. $y = 3(x + 1)$

36. $y = -2(x + 3)$

Summer Income   **Use the following information.**

You earn $15 an hour mowing lawns and $10 an hour washing windows.
You want to make $400 in one week. An algebraic model for your earnings
is $15x + 10y = 400$, where x is the number of hours mowing lawns and
y is the number of hours washing windows.

37. What are your earnings for 3 hours of mowing and 5 hours of
   window washing?

38. Solve the equation for y.

39. Sketch a graph of the equation.

NAME _____ DATE _____

Practice B

For use with pages 210–217

Decide which of the two points lies on the graph of the line.

1. $2x + 4y = 8$ **2.** $3x - y = 8$ **3.** $4y - 3x = -7$

    a. $(2, 1)$ **b.** $(1, 2)$ **a.** $(2, 2)$ **b.** $(3, 1)$ **a.** $(3, 3)$ **b.** $(-1, 1)$

4. $y = 4$ **5.** $x = -3$ **6.** $x = 0$

    a. $(4, 2)$ **b.** $(2, 4)$ **a.** $(-3, 2)$ **b.** $(3, -3)$ **a.** $(0, 3)$ **b.** $(-1, 0)$

7. $y = 4x - 2$ **8.** $y = \frac{1}{2}x + 3$ **9.** $y = -3(x + 1)$

    a. $(-1, -6)$ **b.** $(0, 2)$ **a.** $(-2, 4)$ **b.** $(0, 3)$ **a.** $(-1, -6)$ **b.** $(-2, 3)$

Find three different ordered pairs that are solutions of the equation.

10. $y = 2x + 1$ **11.** $x = 5$ **12.** $y = -4$

13. $y = 5 - 2x$ **14.** $y = 3(2x + 4)$ **15.** $y = -\frac{1}{2}x - 4$

Rewrite the equation in function form.

16. $-2x + y = 6$ **17.** $x + 4y = -2$ **18.** $-x + y = 7$

19. $-5x + 2y = -4$ **20.** $3x - 5y = 1$ **21.** $-2x - 4y = 0$

Use a table of values to graph the equation.

22. $y = 2x + 1$ **23.** $y = 3x - 2$ **24.** $y = -4x + 2$

25. $y = -x - 3$ **26.** $y = \frac{1}{2}x + 3$ **27.** $y = -\frac{1}{4}x + 1$

28. $y = 2$ **29.** $x = -4$ **30.** $y = 0$

31. $y = -(2 - x)$ **32.** $y = -x + \frac{3}{2}$ **33.** $y = -\frac{3}{4}x + \frac{1}{2}$

Summer Income   **Use the following information.**

You earn $15 an hour mowing lawns and $10 an hour washing windows.
You want to make $400 in one week. An algebraic model for your earnings
is $15x + 10y = 400$, where x is the number of hours mowing lawns and
y is the number of hours washing windows.

   34. Solve the equation for y.   **35.** Sketch a graph of the equation.

   36. If you spent 14 hours mowing lawns one week, how many hours did you
      have to wash windows to earn $400?

Distance   **Use the following information.**

You are 455 miles from home and you are driving toward home at a constant
rate of 65 miles per hour. The distance d (in miles) away from home after t
hours is given by $d = 455 - 65t$.

   37. Sketch the graph of the equation from $t = 0$ to $t = 7$.

   38. How far from home are you after 3 hours?

Practice C

For use with pages 210–217

Decide which of the two points lies on the graph of the line.

1. $-2x + y = -6$

  a. $(1, -4)$ **b.** $(-3, 0)$

2. $5x - 3y = 7$

  a. $(2, 0)$ **b.** $(5, 6)$

3. $4y - 6x = 0$

  a. $(-3, -2)$ **b.** $(-2, -3)$

4. $y = 5$

  a. $(5, 2)$ **b.** $(2, 5)$

5. $x = -2$

  a. $(-2, 1)$ **b.** $(2, -2)$

6. $y = 0$

  a. $(0, -3)$ **b.** $(-3, 0)$

7. $y = -x - \frac{3}{4}$

  a. $\left(-1, \frac{1}{4}\right)$ **b.** $\left(\frac{3}{4}, 0\right)$

8. $y = \frac{1}{2}(x + 3)$

  a. $\left(0, \frac{1}{2}\right)$ **b.** $(-1, 1)$

9. $y = -\frac{1}{5}x + \frac{2}{5}$

  a. $(-2, 0)$ **b.** $(7, -1)$

Find three different ordered pairs that are solutions of the equation.

10. $y = 4x + 6$

11. $x = -1$

12. $y = 3$

13. $y = \frac{3}{5}x + 4$

14. $y = -\frac{3}{2}x - \frac{1}{2}$

15. $y = \frac{1}{4}(3x - 1)$

Rewrite the equation in function form.

16. $-7x + y = 1$

17. $-4x + 2y = -6$

18. $-4x - 2y = -1$

19. $3x + 4y = -2$

20. $6x - 5y = 1$

21. $-6x - 9y = 0$

Use a table of values to graph the equation.

22. $y = 1 - 2x$

23. $y = 8x + 5$

24. $y = -\frac{3}{4}x$

25. $y = -\frac{1}{2}x + 3$

26. $y = x - \frac{4}{5}$

27. $y = 3.5$

28. $x = -1.5$

29. $y = -2\left(\frac{1}{3}x + \frac{3}{4}\right)$

30. $y = -\frac{5}{2}x + \frac{3}{2}$

Distance **Use the following information.**

You are 357.5 miles from home and you are driving toward home at a constant rate of 55 miles per hour. The distance, d (in miles), away from home after t hours is given by $d = 357.5 - 55t$.

31. Sketch the graph of the equation from $t = 0$ to $t = 7$.

32. How far from home are you after $3\frac{1}{2}$ hours?

33. After how many hours do you arrive at home?

Burning Calories **Use the following information.**

Sam Jordan burns 10.2 calories per minute mountain biking and 8.6 calories per minute roller blading. His goal is to burn approximately 500 calories daily. An algebraic model for calories burned is $10.2x + 8.6y = 500$, where x is the number of minutes mountain biking and y is the number of minutes roller blading.

34. If Sam only mountain bikes, approximately how many minutes will he need to bike to meet his goal? Round your answer to the nearest minute.

35. If Sam only roller blades, approximately how many minutes will he need to blade to meet his goal? Round your answer to the nearest minute.

36. Use the results of Exercises 34 and 35 to sketch the graph of the equation.

NAME _____ DATE _____

Reteaching with Practice

For use with pages 210–217

GOAL Graph a linear equation using a table or a list of values and graph horizontal and vertical lines.

VOCABULARY

A **solution of an equation** in two variables x and y is an ordered pair (x, y) that makes the equation true.

The **graph of an equation** in x and y is the set of all points (x, y) that are solutions of the equation.

EXAMPLE 1 *Verifying Solutions of an Equation*

Use algebra to decide whether the point $(10, 1)$ lies on the graph of $x - 2y = 8$.

SOLUTION

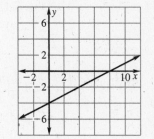

The point $(10, 1)$ appears to be on the graph of $x - 2y = 8$. You can check this algebraically.

$$x - 2y = 8 \qquad \text{Write original equation.}$$
$$10 - 2(1) \overset{?}{=} 8 \qquad \text{Substitute 10 for } x \text{ and 1 for } y.$$
$$8 = 8 \qquad \text{Simplify. True statement}$$

$(10, 1)$ is a solution of the equation $x - 2y = 8$, so it is on the graph.

Exercises for Example 1

Decide whether the given ordered pair is a solution of the equation.

1. $-3x + 6y = 12, (-4, 0)$ **2.** $x + 5y = 11, (2, 1)$

3. $y = 1, (3, 1)$ **4.** $3y - 5x = 4, (-2, 2)$

EXAMPLE 2 *Graphing a Linear Equation*

Use a table of values to graph the equation $x - 2y = 4$.

SOLUTION

Rewrite the equation in function form by solving for y.

$$x - 2y = 4 \qquad \text{Write original equation.}$$
$$-2y = -x + 4 \qquad \text{Subtract } x \text{ from each side.}$$
$$y = \frac{x}{2} - 2 \qquad \text{Divide each side by } -2.$$

NAME _____ DATE _____

Reteaching with Practice

For use with pages 210–217

Choose a variety of values of x and make a table of values.

Choose x.	-4	-2	0	2	4
Evaluate y.	-4	-3	-2	-1	0

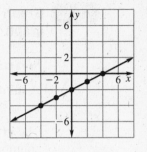

Using the table of values, you can write five ordered pairs.

$(-4, -4), (-2, -3), (0, -2), (2, -1), (4, 0)$

Plot each ordered pair. The line through the points is the graph of the equation.

Exercises for Example 2

Use a table of values to graph the equation.

5. $y = 3x - 4$ **6.** $3y - 3x = 6$ **7.** $y = -3(x - 1)$

EXAMPLE 3 *Graphing y = b*

Graph the equation $y = -3$.

SOLUTION

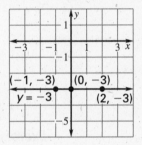

The y-value is always -3, regardless of the value of x. The points $(-1, -3), (0, -3), (2, -3)$ are some solutions of the equation. The graph of the equation is a horizontal line 3 units below the x-axis.

EXAMPLE 4 *Graphing x = a*

Graph the equation $x = 5$.

SOLUTION

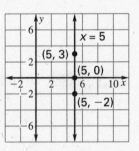

The x-value is always 5, regardless of the value of y. The points $(5, -2), (5, 0), (5, 3)$ are some solutions of the equation. The graph of the equation is a vertical line 5 units to the right of the y-axis.

Exercises for Examples 3 and 4

Graph the equation.

8. $y = 0$ **9.** $x = -4$ **10.** $x = 0$

11. $y = 6$ **12.** $y = -5$ **13.** $x = 2$

Algebra 1
Chapter 4 Resource Book

NAME _____ DATE _____

Quick Catch-Up for Absent Students

For use with pages 210–217

The items checked below were covered in class on (date missed) _____

Lesson 4.2: Graphing Linear Equations

_____ **Goal 1:** Graph a linear equation using a table or a list of values. (pp. 210–211)

Material Covered:

_____ Example 1: Verifying Solutions of an Equation

_____ Student Help: Look Back

_____ Example 2: Graphing an Equation

_____ Student Help: Skills Review

_____ Example 3: Graphing a Linear Equation

_____ Example 4: Using the Graph of a Linear Model

Vocabulary:

solution of an equation, p. 210   graph of an equation, p. 210

_____ **Goal 2:** Graph horizontal and vertical lines. (p. 213)

Material Covered:

_____ Student Help: Study Tip

_____ Example 5: Graphing $y = b$

_____ Example 6: Graphing $x = a$

_____ Other (specify) _____

_____

Homework and Additional Learning Support

_____ Textbook (specify) pp. 214–217 _____

_____

_____ Internet: Extra Examples at www.mcdougallittel.com

_____ *Reteaching with Practice* worksheet (specify exercises)_____

_____ *Personal Student Tutor* for Lesson 4.2

Real Life Application: When Will I Ever Use This?

For use with pages 210–217

Community Service

Community service has become a significant part of high school education. In California, a group called Youth Community Service has been created to connect service to the academic curriculum of nine local schools. Activities include tutoring, service immersion projects, and partnerships with community-based organizations. Some of the benefits for student participants include an increased sense of self-esteem, increased interest in school, cross-cultural opportunities and training, enhanced academic and social skills, and a place to address important issues.

In Exercises 1–6, use the following information.

A school district encourages students to get involved in the community by offering credits for community service hours. The school district offers 0.25 credit for each 45 hours of service. Some examples of how these hours can be earned include volunteering at local senior citizen retirement centers, working with the school's custodial staff, and cleaning up local parks.

Devo'n is a freshman who would like to earn a full credit of service by the time he graduates by volunteering to work at a senior citizen retirement center and with the school's custodial staff. An algebraic model for the number of hours worked during his freshman year is $x + y = 45$, where x is the number of hours he worked at the senior citizen retirement center and y is the number of hours he worked at school with the custodial staff.

1. Rewrite the equation $x + y = 45$ in function form.

2. Use the equation in function form from Exercise 1 to make a table of values for $x = 5$, $x = 15$, $x = 25$, $x = 35$, and $x = 45$.

3. Use the table of values from Exercise 2 to graph the equation.

4. Devo'n ends up working twice as many hours at the senior center than with the school's custodial staff. If he worked a total of 45 hours, how many were spent with the custodial staff?

5. Devo'n decides to spend all of his 45 hours working with the school's custodial staff his sophomore year. Write an equation that models his hours worked.

6. Graph your equation from Exercise 5. What type of line is this?

Algebra 1
Chapter 4 Resource Book

Challenge: Skills and Applications

For use with pages 210–217

1. *Look for a Pattern* **In parts (a)–(d), use a graph to find the coordinates of the point at which the lines cross.**

   a. $x = -4, y = 3$

   b. $x = 1, y = -2$

   c. $x = -5, y = -1$

   d. $x = 0, y = 4$

   e. Write the coordinates of the point at which any lines $x = a$ and $y = b$ cross.

Babysitting **In Exercises 2–7, use the following information.**

For babysitting, you charge $5 an hour plus $2.50 per child for the whole visit.

2. Carry out the following steps to write an algebraic expression for your babysitting rates. Write a verbal model, assign labels, and then translate the verbal model into an algebraic model.

3. Suppose the family for which you do most of your babysitting has three children. How is this situation different from the original information? Write a function that shows how much you would earn babysitting for this family. Describe the input for this function. Describe the output.

4. Make a table of values showing how much you would earn for different lengths of time babysitting with the family in Exercise 3.

5. Use the table of values from Exercise 4 to graph the equation. Use your graph to find out how much to charge for 4 hours of babysitting.

6. Find the value of your function when the number of hours is 0. Does the value make sense in the situation? Explain.

7. Suppose that the family in Exercise 3 has visiting relatives with one child. Will your graph in Exercise 5 still be useful for determining what to charge? Explain.

Algebra 1
Chapter 4 Resource Book

35

TEACHER'S NAME _____ CLASS _____ ROOM _____ DATE _____

Lesson Plan

2-day lesson (See *Pacing the Chapter,* TE pages 200C–200D)   For use with pages 218–224

GOALS
1. **Find the intercepts of the graph of a linear equation.**
2. **Use intercepts to make a quick graph of a linear equation in a real-life problem.**

State/Local Objectives _____

_____

✓ Check the items you wish to use for this lesson.

STARTING OPTIONS
____ Homework Check: TE page 214; Answer Transparencies
____ Warm-Up or Daily Homework Quiz: TE pages 218 and 217, CRB page 38, or Transparencies

TEACHING OPTIONS
____ Motivating the Lesson: TE page 219
____ Lesson Opener (Activity): CRB page 39 or Transparencies
____ Examples: Day 1: 1–3, SE pages 218–219; Day 2: 4, SE page 220
____ Extra Examples: Day 1: TE page 219 or Transp.; Day 2: TE page 220 or Transp.
____ Closure Question: TE page 220
____ Guided Practice: SE page 221; Day 1: Exs. 1–12; Day 2: Ex. 13

APPLY/HOMEWORK
Homework Assignment
____ Basic   Day 1: 14–40 even, 41–43; Day 2: 44–58 even, 60–63, 69, 75, 80, 85, 87, Quiz 1: 1–16
____ Average   Day 1: 14–40 even, 41–43; Day 2: 44–58 even, 60–63, 69, 75, 80, 85, 87, Quiz 1: 1–16
____ Advanced   Day 1: 14–40 even, 41–43; Day 2: 44–58 even, 60–63, 67–70, 75, 80, 85, 87,
                  Quiz 1: 1–16

Reteaching the Lesson
____ Practice Masters: CRB pages 40–42 (Level A, Level B, Level C)
____ Reteaching with Practice: CRB pages 43–44 or Practice Workbook with Examples
____ Personal Student Tutor

Extending the Lesson
____ Applications (Real-Life): CRB page 46
____ Math & History: SE page 224, CRB page 47, or Internet
____ Challenge: SE page 223: CRB page 48 or Internet

ASSESSMENT OPTIONS
____ Checkpoint Exercises: Day 1: TE page 219 or Transp.; Day 2: TE page 220 or Transp.
____ Daily Homework Quiz (4.3): TE page 223, CRB page 52, or Transparencies
____ Standardized Test Practice: SE page 223: TE page 223; STP Workbook: Transparencies
____ Quiz (4.1–4.3): SE page 224: CRB page 49

Notes _____

_____

_____

TEACHER'S NAME _____ CLASS _____ ROOM _____ DATE _____

Lesson Plan for Block Scheduling

1-day lesson (See *Pacing the Chapter,* TE pages 200C–200D)   For use with pages 218–224

GOALS
1. **Find the intercepts of the graph of a linear equation.**
2. **Use intercepts to make a quick graph of a linear equation in a real-life problem.**

State/Local Objectives _____

_____

_____

✓ **Check the items you wish to use for this lesson.**

STARTING OPTIONS

____ Homework Check: TE page 214; Answer Transparencies
____ Warm-Up or Daily Homework Quiz: TE pages 218 and
          217, CRB page 38, or Transparencies

TEACHING OPTIONS

____ Motivating the Lesson: TE page 219
____ Lesson Opener (Activity): CRB page 39 or Transparencies
____ Examples 1–4: SE pages 218–220
____ Extra Examples: TE pages 219–220 or Transparencies
____ Closure Question: TE page 220
____ Guided Practice Exercises: SE page 221

APPLY/HOMEWORK
Homework Assignment

____ Block Schedule: 14–40 even, 41–43, 44–58 even, 60–63, 69, 75, 80, 85, 87, Quiz 1: 1–16

Reteaching the Lesson

____ Practice Masters: CRB pages 40–42 (Level A, Level B, Level C)
____ Reteaching with Practice: CRB pages 43–44 or Practice Workbook with Examples
____ Personal Student Tutor

Extending the Lesson

____ Applications (Real-Life): CRB page 46
____ Math & History: SE page 224, CRB page 47, or Internet
____ Challenge: SE page 223: CRB page 48 or Internet

ASSESSMENT OPTIONS

____ Checkpoint Exercises: TE pages 219–220 or Transparencies
____ Daily Homework Quiz (4.3): TE page 223, CRB page 52, or Transparencies
____ Standardized Test Practice: SE page 223: TE page 223; STP Workbook: Transparencies
____ Quiz (4.1–4.3): SE page 224: CRB page 49

Notes _____

_____

_____

CHAPTER PACING GUIDE	
Day	**Lesson**
1	Assess Ch. 3; 4.1 (all)
2	4.2 (all)
3	**4.3 (all)**
4	4.4 (all)
5	4.5 (all); 4.6 (begin)
6	4.6 (end); 4.7 (all)
7	4.8 (all)
8	Review/Assess Ch. 4

WARM-UP EXERCISES

For use before Lesson 4.3, pages 218–224

1. Which point lies on the *x*-axis, (3, 0) or (0, 3)?

2. Which point lies on the *y*-axis, (−5, 0) or (0, −5)?

The point (*x*, *y*) is a solution of 2*x* + *y* = 8.

3. What is the value of *y* when $x = 0$?

4. What is the value of *x* when $y = 0$?

....

DAILY HOMEWORK QUIZ

For use after Lesson 4.2, pages 210–217

1. Decide whether the given ordered pair is a solution of
$2x - 3y = 8$.

   a. (−2, −4)   **b.** (7, −2)

2. Rewrite $4x - 2y = 18$ in function form.

3. Use a table of values to graph $y = 2x + 2$.

4. Find the coordinates of the point at which the lines cross.

   a. $x = -3, y = 8$ **b.** $y = 0, x = 5$

Lesson 4.3

SET UP: Work with your class.

1. Create a human coordinate plane with members of your class.
Each square on the diagram below shows where one person
should sit.

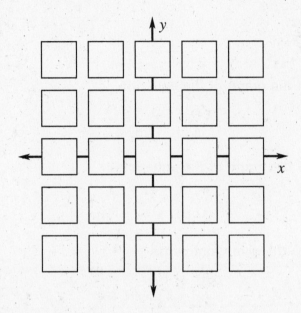

2. Each person on your grid represents a point. What are your
coordinates?

3. Consider the equation $x + y = 1$. Find the coordinates of the
point where the graph of this equation will cross the x-axis.
Stand up if these match your coordinates. Find the coordinates
of the point where the graph of this equation will cross the
y-axis. Stand up if these match your coordinates.

4. Look at the two class members who are standing. If you are not
standing, decide if you are in a position that lines up with the
two students who are standing. If you are, then stand up. If not,
remain seated. What is true about the people standing up?

5. Repeat Steps 3 and 4 for the equation $x - y = 2$.

6. Use your results to make a conjecture about a quick way to
graph a linear equation.

Lesson 4.3

Use the graph to find the *x*-intercept and the *y*-intercept of the line.

1.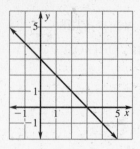

2.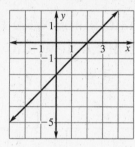

3.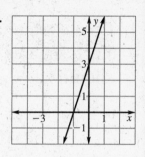

Find the *x*-intercept of the graph of the equation.

4. $x + y = 5$

5. $x - y = -6$

6. $x - 3y = 7$

7. $-3x + y = 15$

8. $2x - 10y = -30$

9. $6x + 12y = 36$

Find the *y*-intercept of the graph of the equation.

10. $y = -3x - 4$

11. $y = \frac{1}{2}x + 6$

12. $y = 3 - 2x$

13. $-3x + 6y = 18$

14. $4x + 4y = -16$

15. $5x - 10y = -40$

Graph the line that has the given intercepts.

16. *x*-intercept: 2
    y-intercept: 2

17. *x*-intercept: 3
    y-intercept: -1

18. *x*-intercept: -3
    y-intercept: 5

19. *x*-intercept: -4
    y-intercept: -5

20. *x*-intercept: -8
    y-intercept: 4

21. *x*-intercept: 10
    y-intercept: -6

Find the *x*-intercept and the *y*-intercept of the line. Graph the equation. Label the points where the line crosses the axes.

22. $y = x + 3$

23. $y = x - 4$

24. $y = 1 + x$

25. $y = 2 - x$

26. $y = 2x - 4$

27. $y = 3x + 5$

28. $-3x + 5y = 15$

29. $-4x + 2y = -4$

30. $7x - 5y = 35$

Ticket Sales   **Use the following information.**

You sold tickets to the school play. Advanced tickets were $4. Tickets bought at the door were $5. Total ticket sales were $400. Let *x* represent the number of advanced tickets sold and *y* represent the number of tickets sold at the door.

31. Graph the linear function $4x + 5y = 400$.

32. Label the *x*-intercept and the *y*-intercept. What does each represent in the situation?

Club Membership   **Use the following information.**

The Spanish Club is open to juniors and seniors. There are now 18 members in the club. Let *x* represent the number of junior members and *y* represent the number of senior members.

33. Graph the linear function $x + y = 18$.

34. Label the *x*-intercept and the *y*-intercept. What does each represent in the situation?

NAME _____ DATE _____

Practice B

For use with pages 218–224

Find the *x*-intercept of the graph of the equation.

1. $x + 2y = 5$
2. $3x - y = 6$
3. $5x + 5y = -30$
4. $6x - 12y = 36$
5. $1.5x - 3y = -6$
6. $0.8x + 3y = 2.4$

Find the *y*-intercept of the graph of the equation.

7. $y = -3x - 7$
8. $y = \frac{1}{2}x + 8$
9. $y = x - \frac{2}{3}$
10. $-3x + 2y = 18$
11. $4x + 2y = -16$
12. $5x - 1.2y = 3.6$

Sketch the line that has the given intercepts.

13. *x*-intercept: 3
    y-intercept: 2
14. *x*-intercept: 4
    y-intercept: -1
15. *x*-intercept: -2
    y-intercept: 5

16. *x*-intercept: -6
    y-intercept: -5
17. *x*-intercept: $\frac{1}{2}$
    y-intercept: -4
18. *x*-intercept: 10
    y-intercept: -6.5

Find the *x*-intercept and the *y*-intercept of the line. Graph the equation. Label the points where the line crosses the axes.

19. $y = x + 6$
20. $y = x - 9$
21. $y = 1 - x$
22. $y = -2 - x$
23. $y = \frac{1}{2}x - 4$
24. $y = -0.5x + 5$
25. $-2x - 4y = 20$
26. $-4x + 8y = -16$
27. $0.3x - 1.3y = 3.9$

Ticket Sales Use the following information.

You sold tickets to the school play. Advanced tickets were $4. Tickets bought at the door were $5.50. Total ticket sales were $440. Let *x* represent the number of advanced tickets sold and *y* represent the number of tickets sold at the door.

28. Write an equation to represent the number of tickets sold.

29. Graph the equation from Exercise 28.

30. What are three possible numbers of advanced tickets sold and tickets sold at the door?

Stacking Crates Use the following information.

As a part of a summer job, you stack crates. The crates have the same length and width, but have heights of 1 or 2 feet. Using a fork lift, you can stack the crates 8 feet high.

31. Make a graph showing the possible number of each type of crate in one stack.

32. If you stacked 3 of the 2-foot crates, how many of the 1-foot crates were in the stack?

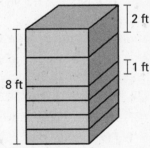

8 ft
2 ft
1 ft

NAME _____ DATE _____

Practice C

For use with pages 218–224

Find the *x*-intercept and the *y*-intercept of the graph of the equation.

1. $x + y = -9$

2. $3x + y = 6$

3. $x - 2y = 4$

4. $-3x + y = -18$

5. $5x + 6y = -30$

6. $-4x + 3y = 6$

In 7–12, sketch the line that has the given intercepts.

7. *x*-intercept: 2
   y-intercept: 5

8. *x*-intercept: 4
   y-intercept: -2

9. *x*-intercept: -1
   y-intercept: 5

10. *x*-intercept: -3
   y-intercept: -5

11. *x*-intercept: -8
   y-intercept: 4

12. *x*-intercept: 10
   y-intercept: -6

Find the *x*-intercept and the *y*-intercept of the line. Graph the equation. Label the points where the line crosses the axes.

13. $y = 7 + x$

14. $y = -6 - 2x$

15. $y = \frac{2}{3}x - 8$

16. $-12x - 2y = 48$

17. $-3x + 9y = -9$

18. $5x - 25y = -50$

19. $y = -0.7x + 7$

20. $-4x - 1.8y = -3.6$

21. $1.4x + 0.7y = -0.28$

Ticket Sales   **Use the following information.**

You sold tickets to the school play. Advanced tickets were $4.50. Tickets bought at the door were $5.50. Total ticket sales were $452. Let *x* represent the number of advanced tickets sold and *y* represent the number of tickets sold at the door.

22. Write an equation to represent the number of tickets sold.

23. Graph the equation from Exercise 22.

24. If you sold 54 advanced tickets, how many tickets were sold at the door?

Carrying Books   **Use the following information.**

You help a teacher move books. The math books weigh 3 pounds each. The science books weigh 4 pounds each. You can carry 24 pounds in one load.

25. Make a graph showing the possible number of each type of book carried in one load.

26. Is it possible for you to carry 5 math books in one load? Explain.

Stacking Crates   **Use the following information.**

As a part of a summer job, you stack crates. The crates have the same length and width, but have heights of $\frac{1}{2}$ foot or $2\frac{1}{2}$ feet. Using a fork lift, you can stack the crates 10 feet high.

27. Make a graph showing the possible number of each type of crate in one stack.

28. If you stacked 2 of the $2\frac{1}{2}$-foot crates, how many of the $\frac{1}{2}$-foot crates were in the stack?

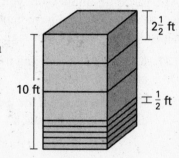

10 ft

$2\frac{1}{2}$ ft

$\frac{1}{2}$ ft

NAME _____ DATE _____

Reteaching with Practice

For use with pages 218–224

GOAL **Find the intercepts of the graph of a linear equation and use the intercepts to sketch a quick graph of a linear equation.**

VOCABULARY

An *x*-intercept is the *x*-coordinate of a point where a graph crosses the *x*-axis. The *y*-coordinate of this point is 0.

A *y*-intercept is the *y*-coordinate of a point where a graph crosses the *y*-axis. The *x*-coordinate of this point is 0.

EXAMPLE 1 *Finding Intercepts*

Find the *x*-intercept and the *y*-intercept of the graph of the equation $4x - 2y = 8$.

SOLUTION

To find the *x*-intercept of $4x - 2y = 8$, let $y = 0$.

$$4x - 2y = 8 \qquad \text{Write original equation.}$$
$$4x - 2(0) = 8 \qquad \text{Substitute 0 for } y.$$
$$x = 2 \qquad \text{Solve for } x.$$

The *x*-intercept is 2. The line crosses the *x*-axis at the point $(2, 0)$.

To find the *y*-intercept of $4x - 2y = 8$, let $x = 0$.

$$4x - 2y = 8 \qquad \text{Write original equation.}$$
$$4(0) - 2y = 8 \qquad \text{Substitute 0 for } x.$$
$$y = -4 \qquad \text{Solve for } y.$$

The *y*-intercept is -4. The line crosses the *y*-axis at the point $(0, -4)$.

Exercises for Example 1

Find the *x*-intercept of the graph of the equation.

1. $x - y = 6$ **2.** $-2x + y = -4$ **3.** $3x - 2y = 6$

Find the *y*-intercept of the graph of the equation.

4. $x - y = 6$ **5.** $-2x + y = -4$ **6.** $3x - 2y = 6$

NAME _____ DATE _____

Reteaching with Practice

For use with pages 218–224

EXAMPLE 2 *Making a Quick Graph*

Graph the equation $2x - y = 8$.

SOLUTION

Find the intercepts by first substituting 0 for y and then substituting 0 for x.

$2x - y = 8$	$2x - y = 8$
$2x - 0 = 8$	$2(0) - y = 8$
$2x = 8$	$-y = 8$
$x = 4$	$y = -8$

The x-intercept is 4.   The y-intercept is -8.

Draw a coordinate plane that includes the points $(4, 0)$ and $(0, -8)$. Plot the points $(4, 0)$ and $(0, -8)$ and draw a line through them. The graph is shown below.

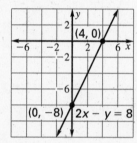

Exercises for Example 2

Find the x-intercept and the y-intercept of the line. Use the intercepts to sketch a quick graph of the equation.

7. $y = -x + 6$ **8.** $x - 5y = 15$ **9.** $y = 4 - 2x$

10. $7x - y = 14$ **11.** $3x + 4y = 24$ **12.** $2y = 7x + 10$

NAME _____ DATE _____

Quick Catch-Up for Absent Students

For use with pages 218–224

The items checked below were covered in class on (date missed) _____

Lesson 4.3: Quick Graphs Using Intercepts

____ **Goal 1:** Find the intercepts of the graph of a linear equation. (p. 218)

Material Covered:

____ Example 1: Find Intercepts

Vocabulary:

    x-intercept, p. 218   *y*-intercept, p. 218

____ **Goal 2:** Use intercepts to make a quick graph of a linear equation in a real-life problem.
      (pp. 219–220)

Material Covered:

____ Example 2: Making a Quick Graph

____ Student Help: Study Tip

____ Example 3: Drawing Appropriate Scales

____ Example 4: Writing and Using a Linear Model

____ Other (specify) _____

_____

Homework and Additional Learning Support

____ Textbook (specify) pp. 221–224 _____

_____

____ *Reteaching with Practice* worksheet (specify exercises) _____

____ *Personal Student Tutor* for Lesson 4.3

Real Life Application:
When Will I Ever Use This?

For use with pages 218–224

Raising Money for a Class Trip

In Exercises 1–7, use the following information.

Your class is selling shirts and sweaters displaying your school logo to raise
money for a field trip. Your class needs to raise $1000 to cover the cost of the
trip. For each shirt sold, $2 is raised for the trip. For each sweater sold, $4 is
raised for the trip.

1. Make a verbal model that represents the number of shirts and sweaters sold
   and the amount of money your class needs to raise for the field trip.

2. Assign labels to your verbal model and write the linear equation. Let x
   represent the number of shirts sold and let y represent the number of
   sweaters sold.

3. Find the x-intercept. What does it represent?

4. Find the y-intercept. What does it represent?

5. A classmate makes a quick graph of your
   equation. The graph is shown at the right.
   Is the graph correct? Why or why not? If
   not, use the x- and y-intercepts to make a
   quick graph.

6. Determine three possible numbers of shirts
   and sweaters to sell that will make your
   class reach its goal.

7. Assume your school has 500 students. How
   many shirts and sweaters do you think your
   class can reasonably expect to sell and still
   reach its goal? Why?

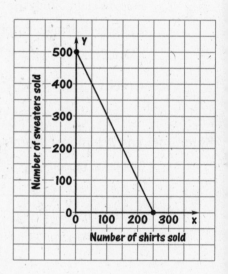

In Exercises 8–12, use the following information.

If your class can raise a total amount of $1500, your class can go on two
field trips.

8. Write a linear equation that represents the numbers of shirts and sweaters
   that your class needs to sell in order to raise enough money for both
   field trips.

9. Find the x- and y-intercepts. What does each intercept represent?

10. Use the x- and y-intercepts to make a quick graph.

11. Determine three possible numbers of shirts and sweaters to sell that will
    make your class reach its goal.

12. Assume your school has 500 students. How many shirts and sweaters
    do you think your class can reasonably expect to sell and still reach its
    goal? Why?

NAME _____ DATE _____

Math and History Application

For use with page 224

HISTORY "It is not enough to have a good mind. The main thing is to use it well." René Descartes wrote these words in 1637. Descartes applied these words to his life because he was a philosopher and scientist, as well as a mathematician. Descartes was born on March 31, 1596, in La Haye, France. At the age of 10 he began school at La Flèche in Anjou and he studied there for 8 years. He studied many subjects including mathematics, classical studies, and philosophy. After graduating, Descartes studied law at the University of Poitiers. Descartes never practiced law, but he did join the military for several years. During his time in the military, Descartes continued to focus his attention toward problems in mathematics and philosophy.

In 1628, Descartes moved to the Netherlands where he spent most of the rest of his life. While in the Netherlands, he wrote many books. One of these books was titled *La Geometrie*. In this book, Descartes changed the way algebraic expressions were written. For example, consider the expression P1 $\mathcal{X} \cdot$ P4 $\mathcal{Z} \cdot$ M7 $\mathcal{C}$. Descartes developed a new system in which this expression is written as $x + 4x^2 - 7x^3$. The system of notation Descartes developed is still used today.

MATH Descartes' contributions to mathematics included applying algebra to geometry and eventually reducing geometric problems to algebraic formulas.
He was interested in studying curves and the relationship between geometry and algebra. Consider the graph of the equation $x^3 + y^3 = 4axy$ when $a = 5$ and when $a = 4$.

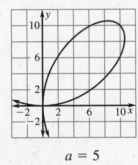

$a = 5$

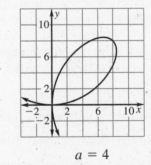

$a = 4$

1. Using the graph of the equation $x^3 + y^3 = 4axy$ when $a = 5$, identify four points on the curve.

2. Using the graph of the equation $x^3 + y^3 = 4axy$ when $a = 4$, identify four points on the curve.

3. Compare the graphs of the equations in Exercises 1 and 2. How are the graphs similar? How are they different?

4. Predict how you think the graph of $x^3 + y^3 = 4axy$ where $a = 3$ will compare with the graphs above. Check your prediction using the graph on page 224 of your textbook.

Lesson 4.3

NAME _____ DATE _____

Challenge: Skills and Applications

For use with pages 218–224

Write an equation of the line with the given *x*- and *y*-intercepts.

Example: *x*-intercept: 6

   y-intercept: 4

Solution: Multiply $6 \cdot 4 = 24$.

   The equation is $4x + 6y = 24$.

   This can be simplified to $2x + 3y = 12$.

   Check: $2(6) + 3(0) = 12$ yes

       $2(0) + 3(4) = 12$ yes

1. *x*-intercept: 3, *y*-intercept: 5

2. *x*-intercept: 8, *y*-intercept: 4

3. *x*-intercept: 9, *y*-intercept: 6

4. *x*-intercept: 7, *y*-intercept: 5

5. **Find a Pattern** Write an equation of the line with *x*-intercept *a* and *y*-intercept *b*.

6. Explain why the first step of the example is to multiply the *x*- and *y*-intercepts. Why does this work?

In Exercises 7–11, use the following information.

Steve is making crafts to sell at a benefit. It takes him $\frac{3}{4}$ of an hour to make a trivet and $\frac{1}{2}$ hour to make a wooden spoon. He has 3 hours to work.

7. Write an equation to show the relationship between how many trivets and how many spoons Steve can make in 3 hours.

8. Graph the function from Exercise 7.

9. What is the *x*-intercept? What does it represent in this situation?

10. What is the *y*-intercept? What does it represent in this situation?

11. What are possible numbers of trivets and spoons Steve can make if he makes at least one of each?

Algebra 1
Chapter 4 Resource Book

NAME _____ DATE _____

Quiz 1

For use after Lessons 4.1–4.3

Answers

1. Plot and label the ordered pairs $A(-6, -2)$, $B(0, -8)$, $C(-3, 2)$, and $D(2, 0)$ in the coordinate plane. *(Lesson 4.1)*

1. Use grid at left. _____

2. Use grid at left. _____

3. _____

_____

4. _____

5. _____

6. _____

7. Use grid at left. _____

For Exercise 1

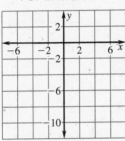

For Exercise 2

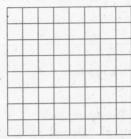

2. The table shows the number of chapters and the numbers of pages for six textbooks. Construct a scatter plot of the data. *(Lesson 4.1)*

Textbooks	English	Math	Science	Health	Social Studies	Spanish
Number of chapters, x	11	14	12	16	23	17
Number of pages, y	252	346	328	310	430	288

3. Find two solutions of the equation $y = 3x - 5$. *(Lesson 4.2)*

4. Rewrite the equation $2x - 4y = 4$ in function form. *(Lesson 4.2)*

5. Write an equation of a horizontal line. *(Lesson 4.2)*

6. Use the graph to find the x-intercept and the y-intercept of the line. *(Lesson 4.3)*

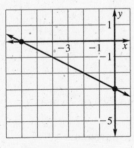

For Exercise 6

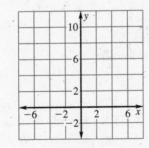

For Exercise 7

7. Find the x-intercept and the y-intercept of $2x + y = 8$. Graph the equation. Label the points where the graph crosses the axes. *(Lesson 4.3)*

Lesson 4.3

TEACHER'S NAME _____ CLASS _____ ROOM _____ DATE _____

Lesson Plan

2-day lesson (See *Pacing the Chapter,* TE pages 200C–200D)   For use with pages 203–209

GOALS 1. **Find the slope of a line using two of its points.**
2. **Interpret slope as a rate of change in a real-life problem.**

State/Local Objectives _____

_____

✓ **Check the items you wish to use for this lesson.**

STARTING OPTIONS

_____ Homework Check: TE page 221; Answer Transparencies
_____ Warm-Up or Daily Homework Quiz: TE pages 226 and 223, CRB page 52, or Transparencies

TEACHING OPTIONS

_____ Motivating the Lesson: TE page 227
_____ Concept Activity: SE page 225; CRB page 53 (Activity Support Master)
_____ Lesson Opener (Visual Approach): CRB page 54 or Transparencies
_____ Examples: Day 1: 1–5, SE pages 226–228; Day 2: 6, SE page 229
_____ Extra Examples: Day 1: TE pages 227–228 or Transp.; Day 2: TE page 229 or Transp.; Internet
_____ Closure Question: TE page 229
_____ Guided Practice: SE page 230; Day 1: Exs. 1–11; Day 2: none

APPLY/HOMEWORK
Homework Assignment

_____ Basic   Day 1: 16–20, 21–35 odd, 36, 39–47 odd; Day 2: 37, 42, 48, 52–59, 67–69, 75, 80, 85, 90, 95
_____ Average   Day 1: 16–20, 21–45 mult. of 3, 47–49; Day 2: 37, 42, 51–61, 67–69, 75, 80, 85, 90, 95
_____ Advanced   Day 1: 16–20, 21–45 mult. of 3, 47–50; Day 2: 51–56, 58–61, 67–71, 75, 80, 85, 90, 95

Reteaching the Lesson

_____ Practice Masters: CRB pages 55–57 (Level A, Level B, Level C)
_____ Reteaching with Practice: CRB pages 58–59 or Practice Workbook with Examples
_____ Personal Student Tutor

Extending the Lesson

_____ Cooperative Learning Activity: CRB page 61
_____ Applications (Interdisciplanary): CRB page 62
_____ Challenge: SE page 233: CRB page 63 or Internet

ASSESSMENT OPTIONS

_____ Checkpoint Exercises: Day 1: TE pages 227–228 or Transp.; Day 2: TE page 229 or Transp.
_____ Daily Homework Quiz (4.4): TE page 233, CRB page 66, or Transparencies
_____ Standardized Test Practice: SE page 233; TE page 233; STP Workbook; Transparencies

Notes _____

_____

_____

TEACHER'S NAME _____ CLASS _____ ROOM _____ DATE _____

Lesson Plan for Block Scheduling

1-day lesson (See *Pacing the Chapter,* TE pages 200C–200D)   For use with pages 225–233

GOALS
1. **Find the slope of a line using two of its points.**
2. **Interpret slope as a rate of change in a real-life problem.**

State/Local Objectives _____

_____

_____

✓ **Check the items you wish to use for this lesson.**

STARTING OPTIONS

____ Homework Check: TE page 221; Answer Transparencies
____ Warm-Up or Daily Homework Quiz: TE pages 226 and
       223, CRB page 52, or Transparencies

TEACHING OPTIONS

____ Motivating the Lesson: TE page 227
____ Concept Activity: SE page 225; CRB page 53 (Activity Support Master)
____ Lesson Opener (Visual Approach): CRB page 54 or Transparencies
____ Examples 1–6: SE pages 226–229
____ Extra Examples: TE pages 227–229 or Transparencies; Internet
____ Closure Question: TE page 229
____ Guided Practice Exercises: SE page 230

APPLY/HOMEWORK

Homework Assignment

____ Block Schedule: 12–48 even, 35, 37, 49, 51–55, 58–60, 67–69, 75, 80, 85, 90, 95

Reteaching the Lesson

____ Practice Masters: CRB pages 55–57 (Level A, Level B, Level C)
____ Reteaching with Practice: CRB pages 58–59 or Practice Workbook with Examples
____ Personal Student Tutor

Extending the Lesson

____ Cooperative Learning Activity: CRB page 61
____ Applications (Interdisciplanary): CRB page 62
____ Challenge: SE page 233; CRB page 63 or Internet

ASSESSMENT OPTIONS

____ Checkpoint Exercises: TE pages 227–229 or Transparencies
____ Daily Homework Quiz (4.4): TE page 233, CRB page 66, or Transparencies
____ Standardized Test Practice: SE page 233; TE page 233; STP Workbook; Transparencies

Notes _____

_____

_____

CHAPTER PACING GUIDE	
Day	Lesson
1	Assess Ch. 3; 4.1 (all)
2	4.2 (all)
3	4.3 (all)
4	**4.4 (all)**
5	4.5 (all); 4.6 (begin)
6	4.6 (end); 4.7 (all)
7	4.8 (all)
8	Review/Assess Ch. 4

NAME _____ DATE _____

WARM-UP EXERCISES
For use before Lesson 4.4, pages 225–233

**Identify the *x*-coordinate and the *y*-coordinate of each
ordered pair.**

1. $(-3, 4)$ **2.** $(0, -7)$

3. $(5, 0)$ **4.** $(-9, -2)$

....

DAILY HOMEWORK QUIZ
For use after Lesson 4.3, pages 218–224

1. Give the *x*- and *y*-intercepts of the graph of $2x - y = -4$.

2. Graph $2x - 3y = 6$. Label the points where the line crosses
the axes.

3. An animal shelter has a total of 60 cats and dogs available
for adoption.

   a. Write an equation to model the situation.

   b. On a graph of the equation the horizontal axis represents the
number of dogs available. What are the intercepts? What do
they represent?

NAME _____ DATE _____

Activity Support Master

For use with page 225

	Height of books	Distance between ruler and books	Rise	Run	Slope
Step 1	7 cm	12 cm	7	12	$\frac{7}{12}$
	11 cm	5.5 cm			
Step 2					
Step 3					
Step 4					
Step 5					

In Exercises 1–3, use the following information.

The slanted side of the triangle shown represents the escalator at the Wilshire/Vermont Metro Rail Station in Los Angeles. You can find the **slope** of this side by writing the following ratio:

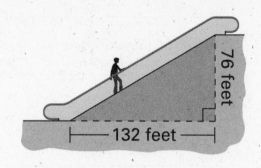

76 feet

132 feet

$$\frac{\text{length of vertical side}}{\text{length of horizontal side}}$$

1. What is the length of the vertical side shown in the diagram?

2. What is the length of the horizontal side shown in the diagram?

3. What is the ratio of the length of the vertical side to the length of the horizontal side? This is the slope of the escalator.

Find the length of the vertical side and the length of the horizontal side shown in each triangle. Then find the slope of the slanted side of the triangle.

4.

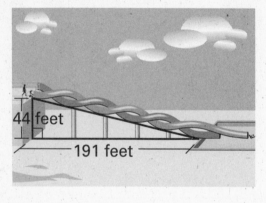

44 feet

191 feet

5.

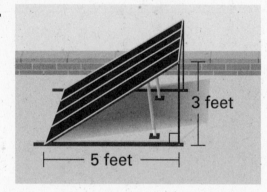

3 feet

5 feet

Practice A

For use with pages 226–233

State whether the slope of the line is *positive, negative, zero,* or *undefined*.

1.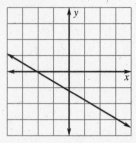

2.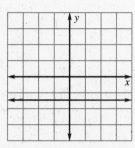

3.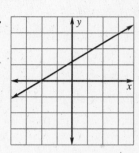

Plot the points and draw a line through them. Without calculating, state whether the slope of the line is *positive, negative, zero,* or *undefined*.

4. $(2, 4), (5, 2)$

5. $(2, -5), (2, 4)$

6. $(4, 1), (6, 7)$

7. $(-3, 5), (2, 5)$

8. $(1, -4), (-2, 3)$

9. $(-4, 2), (0, 5)$

10. $(2, -3), (-4, -3)$

11. $(-5, 1), (5, -1)$

12. $(-1, 3), (-1, -2)$

Find the slope of the line passing through the given points.

13. $(1, 5), (2, 9)$

14. $(2, 4), (1, 1)$

15. $(4, 1), (2, 7)$

16. $(2, 3), (4, 3)$

17. $(0, 4), (-2, 8)$

18. $(6, -8), (6, 4)$

19. $(3, 7), (-9, -5)$

20. $(-2, 3), (4, -1)$

21. $(-5, 2), (2, -4)$

22. $(3, -1), (-6, -1)$

23. $(-3, -9), (-3, -1)$

24. $(-3, -2), (-1, -7)$

Find the value of *y* so that the line passing through the two points has the given slope.

25. $(1, y), (2, 4), m = 1$

26. $(4, y), (5, 3), m = 3$

27. $(-2, 4), (0, y), m = 2$

28. $(3, 5), (1, y), m = -2$

29. $(-2, y), (0, 3), m = -\frac{1}{2}$

30. $(4, -2), (-1, y), m = 1$

31. $(1, 5), (10, y), m = -4$

32. $(-3, 6), (-4, y), m = 5$

33. $(-4, 8), (8, y), m = -9$

In Exercises 34–39, find the rate of change between the two points. Give the units of measure for the rate.

34. $(4, 10)$ and $(6, 15)$; *x* in minutes, *y* in miles

35. $(3, 5)$ and $(11, 69)$; *x* in years, *y* in dollars

36. $(7, 21)$ and $(14, 42)$; *x* in days, *y* in gallons

37. $(1, 2)$ and $(8, 16)$; *x* in weeks, *y* in pounds

38. $(4, 100)$ and $(8, 200)$; *x* in gallons, *y* in miles

39. $(8, 1)$ and $(4, 2)$; *x* in months, *y* in inches

Lesson 4.4

Practice B

For use with pages 226–233

Plot the points and draw a line through them. Without calculating, state whether the slope of the line is *positive, negative, zero*, or *undefined*.

1. $(1, 5), (4, 3)$ **2.** $(-5, 2), (-5, 4)$ **3.** $(3, 3), (7, 6)$

4. $(2, 4), (-3, 4)$ **5.** $(2, -4), (-3, 2)$ **6.** $(-6, 1), (0, 3)$

Find the slope of the line passing through the given points.

7. $(0, 4), (1, 10)$ **8.** $(3, 2), (2, 3)$ **9.** $(5, 2), (3, 8)$

10. $(4, 6), (-2, 6)$ **11.** $(2, 0), (1, 5)$ **12.** $(3, -9), (3, 8)$

13. $(2, 9); (-6, -7)$ **14.** $(-1, 4), (3, -2)$ **15.** $(7, 2), (-8, -3)$

16. $(4, -2), (-8, -2)$ **17.** $(-9, 0), (-9, 7)$ **18.** $(-5, -4), (-3, -9)$

Find the value of *y* so that the line passing through the two points has the given slope.

19. $(2, y), (3, 3), m = 2$ **20.** $(4, y), (6, 3), m = -2$ **21.** $(-3, 5), (0, y), m = 3$

22. $(3, 5), (1, y), m = \frac{3}{2}$ **23.** $(-6, y), (0, 2), m = -\frac{1}{3}$ **24.** $(5, -1), (-2, y), m = 1$

In Exercises 25 and 26, find the rate of change between the two points. Give the units of measure for the rate.

25. $(2, 20)$ and $(4, 42)$; *x* in seconds, *y* in feet

26. $(1, 14)$ and $(3, 40)$; *x* in weeks, *y* in dollars

27. *Postage* In 1989 a postage stamp cost $.25. In 1999 a postage stamp cost $.33. Find the average rate of change in postage in cents per year.

28. *Calculators* In 1975 a 4-function calculator cost $125. In 1995 a 4-function calculator cost $5. Find the average rate of change in the cost of calculators in dollars per year.

Baseball **In Exercises 29–32, use the following information.**

The table shows the number of home runs in major league baseball from 1990 to 1996.

Year	1990	1991	1992	1993	1994	1995	1996
Home runs	3317	3383	3038	4030	3306	4081	4962

29. Calculate the average yearly rate of change in home runs hit from 1990 to 1992.

30. Calculate the average yearly rate of change in home runs hit from 1992 to 1994.

31. Calculate the average yearly rate of change in home runs hit from 1994 to 1996.

32. *Extension* Write a sentence comparing the results of Exercises 29–31 to the average yearly rate of change in home runs hit from 1990 to 1996.

Practice C

For use with pages 226–233

Find the slope of the line passing through the given points.

1. $(1, 2), (9, 6)$ 2. $(-5, -5), (5, 4)$ 3. $(5, 2), (3, 4)$

4. $(4, 6), (5, 4)$ 5. $(-2, 0), (3, 5)$ 6. $(-1, -6), (-1, 3)$

7. $(-4, 4), (-1, -5)$ 8. $(-3, 7), (3, 9)$ 9. $(-5, -4), (5, 6)$

10. $\left(\frac{1}{2}, 2\right), \left(\frac{3}{4}, -1\right)$ 11. $(0.2, 9), (0.2, 5)$ 12. $(-3.3, -2.4), (-1, -7)$

Find the value of *y* so that the line passing through the two points has the given slope.

13. $(1, y), (4, 1), m = 3$ 14. $(3, y), (4, -1), m = -4$ 15. $(-6, 0), (2, y), m = 5$

16. $(-1, y), (5, 2), m = \frac{2}{3}$ 17. $(1, 3), (3, y), m = -\frac{3}{4}$ 18. $(-4, y), (2, 4), m = \frac{1}{2}$

In Exercises 19–22, find the rate of change between the two points. Give the units of measure for the rate.

19. $(2, 150)$ and $(5, 420)$; *x* in hours, *y* in kilometers

20. $(3, 200)$ and $(7, 840)$; *x* in months, *y* in dollars

21. $(1, 10.2)$ and $(9, 90.2)$; *x* in seconds, *y* in meters

22. $(2, 7)$ and $(6, 19)$; *x* in minutes, *y* in inches

23. *Library Books* In 1995 a public library had 16,000 books on its shelves. In 1999 the library had 19,000 books. Find the average rate of change in the number of books per year.

24. *Birth Rate* At Memorial Hospital there were 600 births in 1998. In 2000, there were 550 births. Find the average rate of change in number of births per year.

25. *High School Enrollment* In 1990, about 12.5 million students were enrolled in U.S. high schools. In 2000, about 14.9 million students were projected to be enrolled in U.S. high schools. Find the average rate of change in the number of students per year.

26. *College Enrollment* In 1990, about 13.8 million students were enrolled in U.S. colleges. In 2000, about 14.9 million students were projected to be enrolled in U.S. colleges. Find the average rate of change in the number of students per year.

27. *Voters* About 104 million voters participated in the 1992 U.S. federal election. In 1996, about 96 million voters participated in the U.S. federal election. Find the average rate of change in the number of voters per year.

28. *Unemployment* About 9.6 million U.S. laborers were unemployed in 1992. In 1997, about 6.7 million U.S. laborers were unemployed. Find the average rate of change in the number of unemployed laborers per year.

29. *Extension* Write an expression for the slope of a line passing through the points $(0, -2)$ and $(x, 5)$. What value of *x* will make the fraction equivalent to 1?

30. *Extension* Write an expression for the slope of a line passing through the points $(0, 8)$ and $(x, 2)$. What value of *x* will make the fraction equivalent to -3?

Algebra 1
Chapter 4 Resource Book

Lesson 4.4

LESSON 4.4

Reteaching with Practice

For use with pages 226–233

GOAL Find the slope of a line using two of its points and how to interpret slope as a rate of change in real-life situations.

> **VOCABULARY**
>
> The **slope** m of a nonvertical line is the number of units the line rises or falls for each unit of horizontal change from left to right.
>
> A **rate of change** compares two different quantities that are changing.

EXAMPLE 1 *Finding the Slope of a Line*

Find the slope of the line passing through $(-3, 2)$ and $(1, 5)$.

SOLUTION

Let $(x_1, y_1) = (-3, 2)$ and $(x_2, y_2) = (1, 5)$.

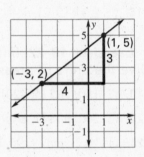

$$m = \frac{y_2 - y_1}{x_2 - x_1} \quad \leftarrow \text{ Rise: Difference of } y\text{-values}$$
$$\qquad\qquad \leftarrow \text{ Run: Difference of } x\text{-values}$$

$$= \frac{5 - 2}{1 - (-3)} \qquad \text{Substitute values.}$$

$$= \frac{3}{1 + 3} = \frac{3}{4} \qquad \text{Simplify. Slope is positive.}$$

Because the slope in Example 1 is positive, the line rises from left to right. If a line has negative slope, then the line falls from left to right.

Exercises for Example 1

Plot the points and find the slope of the line passing through them.

1. $(-4, 0), (3, 3)$ **2.** $(-1, -2), (2, -6)$ **3.** $(-3, -1), (1, 3)$

EXAMPLE 2 *Finding the Slope of a Line*

Find the slope of the line passing through $(-4, 2)$ and $(1, 2)$.

SOLUTION

Let $(x_1, y_1) = (-4, 2)$ and $(x_2, y_2) = (1, 2)$.

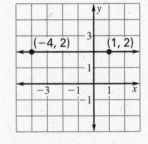

$$m = \frac{y_2 - y_1}{x_2 - x_1} \quad \leftarrow \text{ Rise: Difference of } y\text{-values}$$
$$\qquad\qquad \leftarrow \text{ Run: Difference of } x\text{-values}$$

$$= \frac{2 - 2}{1 - (-4)} \qquad \text{Substitute values.}$$

$$= \frac{0}{5} = 0 \qquad \text{Simplify. Slope is zero.}$$

Lesson 4.4

Reteaching with Practice

For use with pages 226–233

Because the slope in Example 2 is zero, the line is horizontal. If the slope of
a line is undefined, the line is vertical.

Exercises for Example 2

**Plot the points and find the slope of the line passing through
the points.**

4. $(-4, 0), (-4, 3)$ **5.** $(1, -1), (1, 3)$ **6.** $(-3, 0), (1, 0)$

7. $(-4, 3), (1, 3)$ **8.** $(2, -2), (2, -6)$ **9.** $(-1, -6), (2, -6)$

EXAMPLE 3 *Interpreting Slope as a Rate of Change*

In 1994, a video store had 23,500 rentals. In 2000, the store had 28,540
rentals. Find the average rate of change of the store's rentals in rentals
per year.

SOLUTION

Use the formula for slope to find the average rate of change. The change in
rentals is $28,540 - 23,500 = 5040$ rentals. Subtract in the same order. The
change in time is $2000 - 1994 = 6$ years.

VERBAL MODEL	$\boxed{\text{Average rate of change}} = \dfrac{\boxed{\text{Change in rentals}}}{\boxed{\text{Change in time}}}$
LABELS	Average rate of change $= m$ (rentals per year) Change in rentals $= 5040$ (rentals) Change in time $= 6$ (years)
ALGEBRAIC MODEL	$m = \dfrac{5040}{6}$

The average rate of change is 840 rentals per year.

Exercises for Example 3

10. In 1992, the population of Seoul, South Korea was 17,334,000. In 1995,
the population of Seoul was 19,065,000. Find the average rate of
change of the population in people per year.

11. In 1990, the number of motorcycles registered in the United States was
4.3 million. In 1996, the number of registered motorcycles was 3.8 million. Find the average rate of change of the number of registered motorcycles in motorcycles per year.

NAME _____ DATE _____

Quick Catch-Up for Absent Students

For use with pages 225–233

The items checked below were covered in class on (date missed) _____

Activity 4.4: Investigating Slope (p. 225)

____ **Goal:** Use numbers to describe the steepness of a ramp.

Lesson 4.4: The Slope of a Line

____ **Goal 1:** Find the slope of a line using two of its points. (pp. 226–228)

Material Covered:

     ____ Student Help: Look Back

     ____ Example 1: A Line with a Positive Slope Rises

     ____ Example 2: A Line with a Zero Slope is Horizontal

     ____ Student Help: Skills Review

     ____ Example 3: A Line with a Negative Slope Falls

     ____ Example 4: Slope of a Vertical Line is Undefined

     ____ Example 5: Given the Slope, Find a y-Coordinate

Vocabulary:

     slope, p. 226

____ **Goal 2:** Interpret slope as a rate of change in a real-life problem. (p. 229)

Material Covered:

     ____ Example 6: Slope as a Rate of Change

Vocabulary:

     rate of change, p. 229

____ Other (specify) _____

_____

Homework and Additional Learning Support

     ____ Textbook (specify) pp. 230–233_____

_____

     ____ Internet: Extra Examples at www.mcdougallittel.com

     ____ *Reteaching with Practice* worksheet (specify exercises)_____

     ____ *Personal Student Tutor* for Lesson 4.4

Lesson 4.4

Algebra 1
Chapter 4 Resource Book

NAME _____ DATE _____

Cooperative Learning Activity

For use with pages 226–233

GOAL To investigate local wheelchair ramps to see if they meet federal regulations

Materials: wheelchair ramp, yardstick or tape measure, graph paper

Exploring Slope

The American National Standards Institute (ANSI) regulates the construction of new buildings. The Americans with Disabilities Act requires all new public buildings to have wheelchair ramps. The ANSI specifies that the maximum slope of these ramps is 1:12.

Instructions

❶ Find a wheelchair ramp outside your school building. If your school has no ramp, find the nearest ramp in the neighborhood.

❷ Measure the rise and run of the ramp using a yardstick or a tape measure. It may be helpful to measure the rise by measuring the height of the curb.

❸ Make a scale drawing of the ramp on graph paper.

❹ Calculate the slope of the ramp.

Analyzing the Results

1. Does the slope of the ramp meet the ANSI maximum slope standards?

2. What changes could be made to the ramp to decrease the slope? Would those changes be practical for the area surrounding the ramp?

3. Why do you think the ANSI created slope requirements for wheelchair ramps?

Algebra 1
Chapter 4 Resource Book

NAME _____ DATE _____

Interdisciplinary Application

For use with pages 226–233

Minimum Wage

HISTORY The smallest amount of money an employer may legally pay a worker per hour is called a minimum wage. In 1938, President Franklin D. Roosevelt signed the Fair Labor Standards Act. At the time, the act was limited to a few industries and only affected about one-fifth of the labor force. The act established in these industries a minimum wage of 25 cents per hour, banned child labor, and set other standards. Through the years, the act has been amended several times to cover more workers and raise the minimum wage. The minimum wage has grown quickly in recent years. For example, the per hour minimum wage was $1.00 in 1956, $2.00 in 1974, $3.10 in 1980, $4.25 in 1991, and $5.15 in 1997. It should be noted that when a state requires a higher minimum wage than the federal standard, the worker is paid the state minimum wage.

In Exercises 1–4, use the graph at the right.

1. Estimate the average rate of change in the minimum wage from 1955 to 1995 in dollars per year.

2. Estimate the average rate of change in the minimum wage from 1990 to 1997 in dollars per year.

3. Which five-year period had the biggest wage increase?

4. Use the graph to estimate the minimum wage in 2000. Compare your estimate with the actual minimum wage in 2000. Why might your estimate be different from the actual wage?

5. The table below shows the value of the minimum wage from 1955 to 1997 in 1996 dollars. Make a scatter plot of the data.

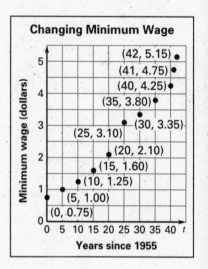

Changing Minimum Wage

Minimum wage (in 1996 Dollars)

Year	1955	1956	1957	1958	1959	1960	1961	1962	1963
Value	$4.39	$5.77	$5.58	$5.43	$5.39	$5.30	$6.03	$5.97	$6.41

Year	1964	1965	1966	1967	1968	1969	1970	1971	1972
Value	$6.33	$6.23	$6.05	$6.58	$7.21	$6.84	$6.47	$6.20	$6.01

Year	1973	1974	1975	1976	1977	1978	1979	1980	1981
Value	$5.65	$6.37	$6.12	$6.34	$5.95	$6.38	$6.27	$5.90	$5.78

Year	1982	1983	1984	1985	1986	1987	1988	1989	1990
Value	$5.45	$5.28	$5.06	$4.88	$4.80	$4.63	$4.44	$4.24	$4.56

Year	1991	1992	1993	1994	1995	1996	1997
Value	$4.90	$4.75	$4.61	$4.50	$4.38	$4.75	$5.03

NAME _____ DATE _____

Challenge: Skills and Applications

For use with pages 226–233

In Exercises 1–4, find the slope of the line passing through the pair of points. Assume a and b are nonzero real numbers.

1. (a, b) and $(2, 5)$

2. (a, b) and $(a, 5)$

3. (a, b) and $(2, b)$

4. (a, b) and (b, a)

5. Name one set of values for a and b so that the line passing through the points (a, b) and $(2, 5)$ has a positive slope.

6. Name one set of values for a and b so that the line passing through the points (a, b) and $(2, 5)$ has a negative slope.

7. If $b < 5$, what must be true about a so that the line passing through the points (a, b) and $(2, 5)$ has a positive slope?

8. If $b < 5$, what must be true about a so that the line passing through the points (a, b) and $(2, 5)$ has a negative slope?

Find the value of a so that the line through the first pair of points is *perpendicular* to the line through the second pair of points. Two nonvertical lines are perpendicular if and only if the slope of one line is the negative reciprocal of the slope of the other line.

Example: $(3, 7)$ and $(5, 4)$; $(4, -2)$ and $(a, -4)$

Solution: slope of line through $(3, 7)$ and $(5, 4)$ $= \dfrac{4 - 7}{5 - 3} = \dfrac{-3}{2}$

slope of perpendicular line $=$ negative reciprocal of $\dfrac{-3}{2} = \dfrac{2}{3}$

So $\dfrac{-4 - (-2)}{a - 4} = \dfrac{2}{3}$

$\dfrac{-2}{a - 4} = \dfrac{2}{3}$

$2(a - 4) = -6$

$a = 1$

9. $(8, 7)$ and $(9, 2)$; $(-1, 3)$ and $(4, a)$

10. $(-2, -3)$ and $(1, 5)$; $(5, -1)$ and $(-3, a)$

11. $(9, -7)$ and $(4, 3)$; $(-2, 8)$ and $(a, 5)$

12. $(1, -4)$ and $(3, 2)$; $(6, 7)$ and $(a, 8)$

Algebra 1
Chapter 4 Resource Book

LESSON 4.5

TEACHER'S NAME _____ CLASS _____ ROOM _____ DATE _____

Lesson Plan

1-day lesson (See *Pacing the Chapter,* TE pages 200C–200D)   **For use with pages 234–239**

 GOALS
1. **Write linear equations that represent direct variation.**
2. **Use a ratio to write an equation for direct variation in a real-life problem.**

State/Local Objectives _____

_____

✓ **Check the items you wish to use for this lesson.**

STARTING OPTIONS

____ Homework Check: TE page 230; Answer Transparencies
____ Warm-Up or Daily Homework Quiz: TE pages 234 and 233, CRB page 66, or Transparencies

TEACHING OPTIONS

____ Motivating the Lesson: TE page 235
____ Lesson Opener (Activity): CRB page 67 or Transparencies
____ Examples 1–4: SE pages 234–236
____ Extra Examples: TE pages 235–236 or Transparencies
____ Closure Question: TE page 236
____ Guided Practice Exercises: SE page 237

APPLY/HOMEWORK
Homework Assignment

____ Basic   12–20 even, 21–31, 33, 38–41, 44–62 even
____ Average   12–20 even, 21–31, 33, 38–41, 44–62 even
____ Advanced   12–20 even, 21–31, 33, 36–43, 44–62 even

Reteaching the Lesson

____ Practice Masters: CRB pages 68–70 (Level A, Level B, Level C)
____ Reteaching with Practice: CRB pages 71–72 or Practice Workbook with Examples
____ Personal Student Tutor

Extending the Lesson

____ Applications (Real-Life): CRB page 74
____ Challenge: SE page 239: CRB page 75 or Internet

ASSESSMENT OPTIONS

____ Checkpoint Exercises: TE pages 235–236 or Transparencies
____ Daily Homework Quiz (4.5): TE page 239, CRB page 78, or Transparencies
____ Standardized Test Practice: SE page 239; TE page 239; STP Workbook; Transparencies

Notes _____

_____

_____

Lesson Plan for Block Scheduling

Half-day lesson (See *Pacing the Chapter,* TE pages 200C–200D)   For use with pages 234–239

GOALS   1. Write linear equations that represent direct variation.
2. Use a ratio to write an equation for direct variation in a real-life problem.

State/Local Objectives _____

_____

_____

✓ **Check the items you wish to use for this lesson.**

STARTING OPTIONS

____ Homework Check: TE page 230; Answer Transparencies
____ Warm-Up or Daily Homework Quiz: TE pages 234 and
        233, CRB page 66, or Transparencies

TEACHING OPTIONS

____ Motivating the Lesson: TE page 235
____ Lesson Opener (Activity): CRB page 67 or Transparencies
____ Examples 1–4: SE pages 234–236
____ Extra Examples: TE pages 235–236 or Transparencies
____ Closure Question: TE page 236
____ Guided Practice Exercises: SE page 237

APPLY/HOMEWORK
Homework Assignment (See also the assignment for Lesson 4.6.)
____ Block Schedule: 12–20 even, 21–31, 33, 38–41, 44–62 even

Reteaching the Lesson

____ Practice Masters: CRB pages 68–70 (Level A, Level B, Level C)
____ Reteaching with Practice: CRB pages 71–72 or Practice Workbook with Examples
____ Personal Student Tutor

Extending the Lesson

____ Applications (Real-Life): CRB page 74
____ Challenge: SE page 239; CRB page 75 or Internet

ASSESSMENT OPTIONS

____ Checkpoint Exercises: TE pages 235–236 or Transparencies
____ Daily Homework Quiz (4.5): TE page 239, CRB page 78, or Transparencies
____ Standardized Test Practice: SE page 239; TE page 239; STP Workbook; Transparencies

Notes _____

_____

_____

_____

CHAPTER PACING GUIDE	
Day	**Lesson**
1	Assess Ch. 3; 4.1 (all)
2	4.2 (all)
3	4.3 (all)
4	4.4 (all)
5	**4.5 (all)**; 4.6 (begin)
6	4.6 (end); 4.7 (all)
7	4.8 (all)
8	Review/Assess Ch. 4

NAME _____ DATE _____

WARM-UP EXERCISES

For use before Lesson 4.5, pages 234–239

Identify the slope and *y*-intercept of the graph of each equation.

1. $y = 2x$

2. $-9x = 3y$

3. You are driving to a national park for a family vacation. If you travel at 60 mi/h, write an equation that describes the relationship between the number of miles *d* you drive and the number of hours *t* you spend driving.

DAILY HOMEWORK QUIZ

For use after Lesson 4.4, pages 225–233

1. Plot the points and draw the line through them. Calculate the slope of the line.

a. $(-2, 2), (0, 4)$ **b.** $(1, 1), (4, 2)$

2. The points $(-4, -2), (0, 0),$ and $(4, y)$ are on the same line. What is the value of *y*?

3. Find the rate of change between the given points. Give the units of measure for the rate. $(4, 24)$ and $(20, 120)$, *x* in hours, *y* in dollars

NAME ———————————————— DATE ————

Activity Lesson Opener

For use with pages 234–239

SET UP: Work in a group.

YOU WILL NEED: • ruler   • your algebra textbook

1. Measure the thickness of the spine of your algebra book.

2. Suppose you stack one algebra book on top of another. What is the height of the stack of two books?

3. Suppose you make a stack of three algebra books. What is the height of the stack of three books?

4. Use the coordinate grid below. Fill in the units you used for measuring height, choose an appropriate scale, and label the y-axis. Then use your answers to Steps 1–3 to plot points to show the height y of x algebra books.

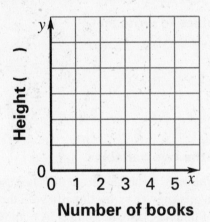

Number of books

5. Draw a line through the points. Write the equation of the line.

6. You have shown that the height y of a stack of algebra books varies directly with the number x of books. The model for direct variation has the form $y = kx$, where k is the constant of variation. What is the constant of variation for this equation? What is the meaning of the constant in this equation?

LESSON
4.5

Practice A

For use with pages 234–239

Determine if the graph represents a direct variation model. If yes, find the constant of variation and the slope.

1. $y = x$

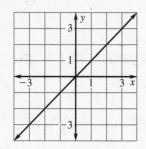

2. $y = x - 1$

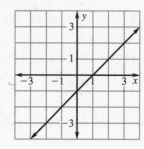

3. $y = -3x$

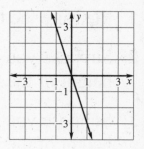

Graph the equation. Find the constant of variation and the slope of the direct variation model.

4. $y = 2x$

5. $y = 3x$

6. $y = 4x$

7. $y = -x$

8. $y = -2x$

9. $y = \frac{1}{3}x$

The variables *x* and *y* vary directly. Use the given values to write an equation that relates *x* and *y*.

10. $x = 1, y = 2$

11. $x = 1, y = 3$

12. $x = 2, y = 4$

13. $x = 3, y = 12$

14. $x = 2, y = -2$

15. $x = -3, y = 15$

16. $x = -4, y = -28$

17. $x = 2, y = 1$

18. $x = 4, y = 1$

In Exercises 19 and 20, find an equation that relates the two variables.

19. *Circumference and Radius* The circumference C of a circle varies directly with the length of the radius r. When the circumference is 8π, the radius is 4.

20. *Showers* The gallons G of water used to take a shower varies directly with the number of minutes M in the shower. A 6 minute shower uses 36 gallons of water.

Salary **Use the following information.**

You work a different number of hours each day. The table shows your total pay p and the number of hours h you worked.

21. Complete the table by finding the ratio of your total pay each day to the number of hours you worked that day.

Total pay, p	$20	$15	$30	$25
Hours worked, h	4	3	6	5
Ratio				

22. Write the model that relates the variables p and h.

23. If you work 8 hours on the fifth day, what will your total pay be?

NAME _____ DATE _____

Practice B

For use with pages 234–239

Determine if the graph represents a direct variation model. If yes, find the constant of variation and the slope.

1. $y = 5x$

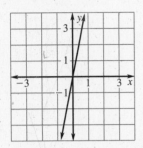

2. $y = -2x + 1$

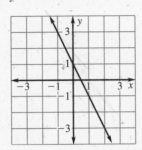

3. $y = \frac{1}{2}x$

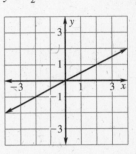

Graph the equation. Find the constant of variation and the slope of the direct variation model.

4. $y = 3x$

5. $y = -4x$

6. $y = 0.5x$

7. $y = -0.2x$

8. $y = -\frac{2}{3}x$

9. $y = \frac{1}{4}x$

The variables x and y vary directly. Use the given values to write an equation that relates x and y.

10. $x = 4, y = 32$

11. $x = 36, y = 9$

12. $x = 15, y = 45$

13. $x = -2, y = 8$

14. $x = 12, y = -3$

15. $x = 10, y = 16$

16. $x = -8, y = 12$

17. $x = 1, y = 0.3$

18. $x = -5, y = -8$

Assume the variables vary directly. Use an equation to find the value of y.

19. If $x = 2$ when $y = 8$, find y when $x = 16$.

20. If $x = 14$ when $y = 7$, find y when $x = 10$.

21. If $x = 3$ when $y = 4$, find y when $x = 24$.

22. If $x = 6$ when $y = 10$, find y when $x = 9$.

In Exercises 23 and 24, find an equation that relates the two variables. Then solve the problem.

23. *Circumference and Radius*   The circumference C of a circle varies directly with the length of the radius r. When the circumference is 15π, the radius is 7.5. Find the circumference when $r = 2.5$.

24. *Salary and Hours*   The salary s an hourly employee earns varies directly with the number of hours h worked. When the salary is $196.80, the hours worked is 24. Find the salary for a 40 hour work week.

25. *Showers*   The gallons G of water used to take a shower varies directly with the number of minutes M in the shower. The standard shower head uses about 6 gallons of water per minute. Write an equation that relates M and G. Find the value of G when $M = 8.5$.

26. *Hooke's Law*   The force F required to stretch a spring varies directly with the amount the spring is stretched s. Ten pounds is needed to stretch a spring 10 inches. Find the force required to stretch the spring 4 inches.

Practice C

For use with pages 234–239

Determine if the graph represents a direct variation model. If yes, find the constant of variation and the slope.

1. $y = -\frac{1}{3}x$

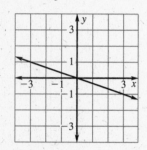

2. $y = -\frac{2}{3}x + 6$

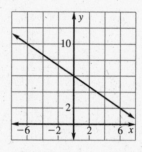

3. $y - 5x = 0$

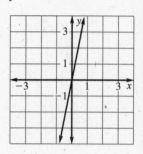

Graph the equation. Find the constant of variation and the slope of the direct variation model.

4. $y = 6x$　　　　　　**5.** $y = -10x$　　　　　**6.** $y = -0.6x$

7. $y = 1.2x$　　　　　**8.** $y = \frac{1}{6}x$　　　　　**9.** $y = -\frac{4}{3}x$

The variables x and y vary directly. Use the given values to write an equation that relates x and y.

10. $x = 3, y = 27$　　　　　**11.** $x = -4, y = 44$　　　　　**12.** $x = -3, y = -21$

13. $x = -3, y = 2$　　　　　**14.** $x = 15, y = -6$　　　　　**15.** $x = -20, y = -24$

16. $x = -50, y = 75$　　　　　**17.** $x = 1, y = 0.9$　　　　　**18.** $x = -0.6, y = 1.2$

Assume the variables vary directly. Use an equation to find the value of x.

19. If $x = 4$ when $y = 8$, find x when $y = 8$.　　　**20.** If $x = -2$ when $y = 14$, find x when $y = 21$.

21. If $x = 4$ when $y = 3$, find x when $y = 12$.　　　**22.** If $x = 16$ when $y = -18$, find x when $y = 36$.

Find an equation that relates the two variables. Then solve the problem.

23. *Ohm's Law*　The electromotive force, E (in volts), varies directly with the current, I (in amperes), flowing through a given conductor. A current of 4 amperes passing through the conductor has an electromotive force of 8 volts. What is the electromotive force if 3 amperes pass through the conductor?

24. *Circumference and Radius*　The circumference C of a circle varies directly with the length of the radius r. When the circumference is $\frac{1}{2}\pi$, the radius is $\frac{1}{4}$. Find the radius when the circumference is 10π.

25. *Showers*　The gallons G of water used to take a shower varies directly with the number of minutes M in the shower. A 6.5 minute shower uses 39 gallons of water. Find how long it would take to use 60 gallons of water.

26. *Hooke's Law*　The force F required to stretch a spring varies directly with the amount the spring is stretched s. Ten pounds is needed to stretch a spring 8 inches. Find the amount the spring is stretched when the force is 20 pounds.

Algebra 1
Chapter 4 Resource Book

NAME _____ DATE _____

Reteaching with Practice

For use with pages 234–239

GOAL Write linear equations that represent direct variation and use a ratio to write an equation for direct variation.

VOCABULARY

In the model for direct variation $y = kx$, the nonzero number k is the **constant of variation.**

Two quantities that vary directly are said to have **direct variation.**

EXAMPLE 1 *Writing a Direct Variation Equation*

The variables x and y vary directly. When $x = 4$, $y = 6$.

a. Write an equation that relates x and y.

b. Find the value of y when $x = 12$.

SOLUTION

a. Because x and y vary directly, the equation is of the form $y = kx$. You can solve for k as follows.

$y = kx$	Write model for direct variation.
$6 = k(4)$	Substitute 4 for x and 6 for y.
$1.5 = k$	Divide each side by 4.

An equation that relates x and y is $y = 1.5x$.

b. $y = 1.5(12)$ Substitute 12 for x in $y = 1.5x$.
$y = 18$ Simplify.

When $x = 12$, $y = 18$.

Exercises for Example 1

In Exercises 1–6, the variables x and y vary directly. Use the given values to write an equation that relates x and y.

1. $x = 3, y = 15$ **2.** $x = 6, y = 3$ **3.** $x = -4, y = -4$

4. $x = 10, y = -2$ **5.** $x = 3.5, y = 7$ **6.** $x = -12, y = 4$

NAME _____ DATE _____

Reteaching with Practice

For use with pages 234–239

EXAMPLE 2 *Using a Ratio to Write a Model*

Weight varies directly with gravity. A person who weighs 150 pounds on Earth weighs 57 pounds on Mars.

a. Write a model that relates a person's weight E on Earth to that person's weight M on Mars.

b. A person weighs 210 pounds on Earth. Use the model to estimate that person's weight on Mars.

SOLUTION

a. Rewrite the model $E = kM$ for direct variation as $k = \dfrac{E}{M}$.

This is the ratio form of a direct variation model. When $E = 150$ and $M = 57$, $k = \dfrac{150}{57}$. The model for direct variation is $E = \dfrac{150}{57}M$.

b. Use the model $E = \dfrac{150}{57}M$ to estimate the person's weight on Mars.

$210 = \dfrac{150}{57}M$ Substitute 210 for E.

$79.8 \approx M$ Multiply each side by $\dfrac{57}{150}$.

You estimate that the person weighs about 79.8 pounds on Mars.

Exercises for Example 2

7. Use the ratio model $E = \frac{150}{57}M$ to estimate a person's weight on Mars if the person weighs 120 pounds on Earth.

8. Use the ratio model $E = \frac{150}{57}M$ to estimate a person's weight on Earth if the person weighs 62 pounds on Mars.

9. A person who weighs 160 pounds on Earth weighs 139 pounds on Venus.

    a. Write a model that relates a person's weight E on Earth to that person's weight V on Venus.

    b. A person weighs 195 pounds on Earth. Use the model to estimate that person's weight on Venus.

Quick Catch-Up for Absent Students

For use with pages 234–239

The items checked below were covered in class on (date missed) _____

Lesson 4.5: Direct Variation

_____ **Goal 1:** Write linear equations that represent direct variation. (pp. 234–235)

Material Covered:

_____ Example 1: Graphs of Direct Variation Models

_____ Example 2: Writing a Direct Variation Equation

_____ Example 3: Writing a Direct Variation Model

Vocabulary:

constant of variation, p. 234   direct variation, p. 234

_____ **Goal 2:** Use a ratio to write an equation for direct variation in a real-life problem. (p. 236)

Material Covered:

_____ Example 4: Using a Ratio to Write a Model

_____ Other (specify) _____

_____

Homework and Additional Learning Support

_____ Textbook (specify) <u>pp. 237–239</u> _____

_____

_____ *Reteaching with Practice* worksheet (specify exercises)_____

_____ *Personal Student Tutor* for Lesson 4.5

Real Life Application:
When Will I Ever Use This?

For use with pages 234–239

Gasoline Prices

In Sacramento, California, gasoline prices fluctuated dramatically during the first half of 1999. After recording near record lows of $1.05 per gallon in February, fires and mechanical failures that shut down four California refineries drove up prices to around $1.67 per gallon in April. Because of California's strict clean-air specifications set by the California Air Resources Board (CARB), obtaining gas from other refineries was not an option. Wholesale distributors, fearing they would run out of gasoline that met CARB specifications, bid up gasoline prices. After the refineries re-opened, prices once again began falling and dropped to around $1.42 per gallon by May. Increases in worldwide crude oil prices, the main factor in driving gasoline prices up (or down), kept the price of gasoline from returning to the pre-crisis levels.

In Exercises 1–3, use the following information.

The cost of gasoline in dollars at a gas station varies directly with the number of gallons of gasoline that you pump. It costs $17.42 to fill your 13-gallon tank at a station in Sacramento.

1. Write a direct variation model that relates the number of gallons g to the total cost c (in dollars) to fill the tank.

2. Use your model from Exercise 1 to determine how much it will cost to fill-up a car with a 19-gallon tank.

3. If you decide to buy a higher grade of gasoline, what will change in your model?

In Exercises 4 and 5, use the following information.

In many collegiate towns, gasoline stations raise their prices when students return to campus in August. The cost of gasoline (in dollars) and the number of gallons pumped by selected customers in eight university towns in Indiana are shown in the table below.

University	Town	Total Cost	Number of Gallons
Ball State	Muncie	$10.71	9
DePauw	Greencastle	$26.18	22
Indiana State	Terre Haute	$22.61	19
Indiana	Bloomington	$38.08	32
Purdue	West Lafayette	$15.47	13
Taylor	Fort Wayne	$19.04	16
Notre Dame	South Bend	$29.75	25
Valparaiso	Valparaiso	$16.66	14

4. Write a ratio model that relates the total cost for gasoline to the number of gallons pumped.

5. Estimate the total cost for a car that needs 18 gallons of gasoline to fill the tank.

Algebra 1
Chapter 4 Resource Book

LESSON
4.5

Challenge: Skills and Applications

For use with pages 234–239

In Exercises 1–3, state whether the two quantities have direct variation. If they do, find the constant of variation and write an equation that relates the variable quantities.

1. Marisa Margolez fenced a portion of a field. The fenced portion had an area of 88 square yards and a length of 11 yards. Later Marisa decided to keep the width of the fenced portion of the field the same and change the length of the fenced portion to 15 yards. Under these circumstances, do the area (A) and the length (L) have direct variation?

2. Start with the original dimensions for the fenced portion of the field in Exercise 1. Suppose Marisa Margolez decided to keep the area of the fenced portion the same as the original area but change the width of the fenced portion to 4 yards. Under these circumstances, do the length (L) and the width (W) have direct variation?

3. Karl Ivanovic designed a square-bottomed pan with a volume of 2400 cubic centimeters and a depth of 6 centimeters. He decided to keep the dimensions of the bottom of the pan the same and change the depth. Under these circumstances, do the volume of the pan (V) and the depth (d) have direct variation?

4. Suppose Karl Ivanovic from Exercise 3 kept the 6-centimeter depth the same in his pan with a volume of 2400 cubic centimeters and changed the length of the sides of the square bottom. Under these circumstances, is there a relationship between the volume of the pan (V) and the length of a side of the square bottom (s)?

5. Calories burned varies directly with time spent walking. David Wong burned 150 calories walking for 35 minutes. How many calories does he burn when he walks 2 hours in a week?

6. The cost of a piece of lace trim varies directly with the length of the piece. Find the cost of 8 yards of trim if 2 feet cost $2.50.

7. Two points of a function have coordinates (p, $10p$) and ($q + 1$, $10q + 10$). Do these points belong to a direct variation model? If so, what is the constant of variation?

TEACHER'S NAME _____ CLASS _____ ROOM _____ DATE _____

Lesson Plan

2-day lesson (See *Pacing the Chapter,* TE pages 200C–200D)   For use with pages 240–249

 GOALS   1. **Graph a linear equation in slope-intercept form.**
2. **Graph and interpret equations in slope-intercept form in a real-life problem.**

State/Local Objectives _____

_____

✓ Check the items you wish to use for this lesson.

STARTING OPTIONS

_____ Homework Check: TE page 237; Answer Transparencies

_____ Warm-Up or Daily Homework Quiz: TE pages 241 and 239, CRB page 78, or Transparencies

TEACHING OPTIONS

_____ Motivating the Lesson: TE page 242

_____ Concept Activity: SE page 240; CRB page 79 (Activity Support Master)

_____ Lesson Opener (Application): CRB page 80 or Transparencies

_____ Graphing Calculator Activity with Keystrokes: CRB pages 81–85

_____ Examples: Day 1: 1–3, SE pages 241–242; Day 2: 4, SE page 243

_____ Extra Examples: Day 1: TE page 242 or Transp.; Day 2: TE page 243 or Transp.

_____ Technology Activity: SE pages 248–249

_____ Closure Question: TE page 243

_____ Guided Practice: SE page 244; Day 1: Exs. 1–10, Day 2: Exs. 11–12

APPLY/HOMEWORK

Homework Assignment

_____ Basic   Day 1: 14–50 even; Day 2: 56–59, 62–65, 69, 70, 81–89, Quiz 2: 1–19

_____ Average   Day 1: 14–50 even; Day 2: 56–65, 70, 81–89, Quiz 2: 1–19

_____ Advanced   Day 1: 14–50 even; Day 2: 62–77, 81–89, Quiz 2: 1–19

Reteaching the Lesson

_____ Practice Masters: CRB pages 86–88 (Level A, Level B, Level C)

_____ Reteaching with Practice: CRB pages 89–90 or Practice Workbook with Examples

_____ Personal Student Tutor

Extending the Lesson

_____ Applications (Interdisciplinary): CRB page 92

_____ Challenge: SE page 246: CRB page 93 or Internet

ASSESSMENT OPTIONS

_____ Checkpoint Exercises: Day 1: TE page 242 or Transp.; Day 2: TE page 243 or Transp.

_____ Daily Homework Quiz (4.6): TE page 247, CRB page 97, or Transparencies

_____ Standardized Test Practice: SE page 246; TE page 247; STP Workbook; Transparencies

_____ Quiz (4.4–4.6): SE page 247; CRB page 94

Notes _____

_____

_____

LESSON
4.6

TEACHER'S NAME _____ CLASS _____ ROOM _____ DATE _____

Lesson Plan for Block Scheduling

2-day lesson (See *Pacing the Chapter,* TE pages 200C–200D)   For use with pages 240–249

GOALS   1. **Graph a linear equation in slope-intercept form.**
2. **Graph and interpret equations in slope-intercept form in a real-life problem.**

State/Local Objectives _____

_____

_____

Lesson 4.6

CHAPTER PACING GUIDE	
Day	**Lesson**
1	Assess Ch. 3; 4.1 (all)
2	4.2 (all)
3	4.3 (all)
4	4.4 (all)
5	4.5 (all); **4.6 (begin)**
6	**4.6 (end)**; 4.7 (all)
7	4.8 (all)
8	Review/Assess Ch. 4

✓ **Check the items you wish to use for this lesson.**

STARTING OPTIONS

____ Homework Check: TE page 237; Answer Transparencies
____ Warm-Up or Daily Homework Quiz: TE pages 241 and
        239, CRB page 78, or Transparencies

TEACHING OPTIONS

____ Motivating the Lesson: TE page 242
____ Concept Activity: SE page 240; CRB page 79 (Activity Support Master)
____ Lesson Opener (Application): CRB page 80 or Transparencies
____ Graphing Calculator Activity with Keystrokes: CRB pages 81–85
____ Examples: Day 5: 1–3, SE pages 241–242; Day 6: 4, SE page 243
____ Extra Examples: Day 5: TE page 242 or Transp.; Day 6: TE page 243 or Transp.
____ Technology Activity: SE pages 248–249
____ Closure Question: TE page 243
____ Guided Practice: SE page 244; Day 5: Exs. 1–10, Day 6: Exs. 11–12

APPLY/HOMEWORK

Homework Assignment (See also the assignments for Lessons 4.5 and 4.7.)

____ Block Schedule: Day 5: 14–50 even; Day 6: 56–65, 69, 70, 81–89, Quiz 2: 1–19

Reteaching the Lesson

____ Practice Masters: CRB pages 86–88 (Level A, Level B, Level C)
____ Reteaching with Practice: CRB pages 89–90 or Practice Workbook with Examples
____ Personal Student Tutor

Extending the Lesson

____ Applications (Interdisciplinary): CRB page 92
____ Challenge: SE page 246; CRB page 93 or Internet

ASSESSMENT OPTIONS

____ Checkpoint Exercises: Day 5: TE page 242 or Transp.; Day 6: TE page 243 or Transp.
____ Daily Homework Quiz (4.6): TE page 247, CRB page 97, or Transparencies
____ Standardized Test Practice: SE page 246; TE page 247; STP Workbook; Transparencies
____ Quiz (4.4–4.6): SE page 247; CRB page 94

Notes _____

_____

_____

WARM-UP EXERCISES

For use before Lesson 4.6, pages 240–249

Use a graph to identify the slope and *y*-intercept of each equation.

1. $y = -6x$

2. $y = 3x - 7$

3. $4x - y = 5$

4. $7x - 3y = 4$

Lesson 4.6

DAILY HOMEWORK QUIZ

For use after Lesson 4.5, pages 234–239

1. Find the constant of variation and the slope of the direct variation model.

  a. $y = \dfrac{1}{4}x$

  b. $y = -2.5x$

2. The perimeter of a square with side length s is modeled by $p = 4s$. Do the side length and perimeter have direct variation?

3. The variables x and y vary directly. Use the given values to write an equation that relates x and y.

  a. $x = 5, y = 18$

  b. $x = -4, y = 6.8$

4. Sound travels about 12.4 miles in one minute. How long does it take sound to travel 5 miles?

Algebra 1
Chapter 4 Resource Book

NAME _____ DATE _____

Activity Support Master

For use with page 240

Step 1

Thickness of algebra textbook: _____

Height of the top of your desk to the floor: _____

Step 2

Model: _____

Step 3

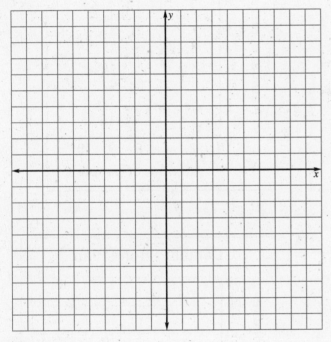

Step 4

Thickness of English textbook: _____

Model: _____

Step 5

Thickness of another book: _____

Model: _____

NAME _____ DATE _____

Application Lesson Opener

For use with pages 241–247

**Give the slope and intercept(s) for each graph shown below.
Give the real-world interpretation of the slope and intercepts.
Hint: Be sure to look carefully at the scale on each axis.**

1.

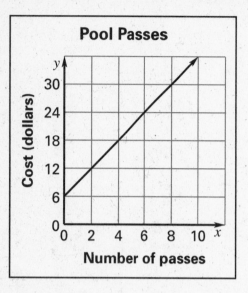

2.

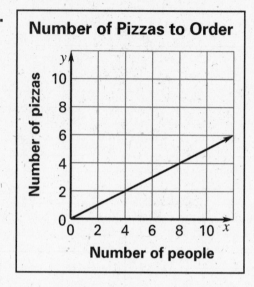

3.

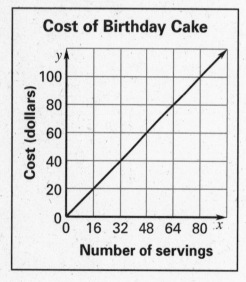

4.

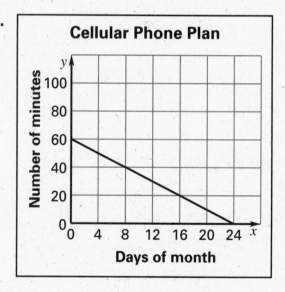

5. Suppose that you drew the graphs above using just the slope
and *y*-intercept of each line. Explain how you could have
drawn the graphs.

Algebra 1
Chapter 4 Resource Book

NAME _____ DATE _____

Graphing Calculator Activity

For use with pages 241–247

GOAL   **To determine whether two different lines in the same plane are parallel.**

Two different lines in the same plane are parallel if they do not intersect. In a coordinate plane, any two vertical lines are parallel. Any two nonvertical lines are parallel if and only if they have the same slope.

A graphing calculator can be used to visually check whether two different lines are parallel. If the lines are not parallel, you can use the graphing calculator's   **TRACE**   and   **ZOOM**   features to determine where the lines intersect.

Activity

❶   Enter each equation into your graphing calculator.

line a: $y = 4$ line b: $y = 4x$ line c: $y = \frac{1}{4}x + 1$

line d: $y = 4x + 4$ line e: $y = -4x + 3$ line f: $y = 4x - 8$

❷   Plot the graph of each equation in the same coordinate plane.

❸   Visually check which lines are parallel. What do their equations have in common?

❹   For lines that are not parallel, determine where the lines intersect using the   **TRACE**   and   **ZOOM**   features.

Exercises

1.   Determine the unknown slope that would make the lines parallel. Then use your graphing calculator to visually check your answer.

a. $y = 3x + 2$ **b.** $y = -5x + 1$ **c.** $y = \frac{2}{3}x - 0.9$

   $y = \underline{\ ?\ } x$ $y = \underline{\ ?\ } x$ $y = \underline{\ ?\ } x$

2.   For each part of Exercise 1, write equations of two other lines that are parallel to the original line. Then use your graphing calculator to visually check your answer.

See page 82 for keystrokes.

Graphing Calculator Activity

For use with pages 241–247

TI-82

Y=	3	ENTER
(-)	1	ENTER
(| 1 | ÷ | 4 |)	X,T,θ	+	1	ENTER
4	X,T,θ	+	4	ENTER
(-)	4	X,T,θ	+	3	ENTER
4	X,T,θ	−	8	ENTER
ZOOM	6

Press **TRACE** and use the left and right arrow keys to move the trace cursor. Move the trace cursor to estimate the point where two lines intersect.

ZOOM 2 **ENTER**

Repeat the trace and zoom steps to get a more accurate estimate.

TI-83

Y=	3	ENTER
(-)	1	ENTER
(| 1 | ÷ | 4 |)	X,T,θ	+	1	ENTER
4	X,T,θ,n	+	4	ENTER
(-)	4	X,T,θ,n	+	3	ENTER
4	X,T,θ,n	−	8	ENTER
ZOOM	6

Press **TRACE** and use the left and right arrow keys to move the trace cursor. Move the trace cursor to estimate the point where two lines intersect.

ZOOM 2 **ENTER**

Repeat the trace and zoom steps to get a more accurate estimate.

Algebra 1
Chapter 4 Resource Book

NAME _____ DATE _____

Graphing Calculator Activity

For use with pages 241–247

SHARP EL-9600c

| Y= | 3 | ENTER |

| (-) | 1 | ENTER |

| (| 1 | ÷ | 4 |) | X/θ/T/n | + | 1 | ENTER |

4 | X/θ/T/n | + | 4 | ENTER |

| (-) | 4 | X/θ/T/n | + | 3 | ENTER |

4 | X/θ/T/n | − | 8 | ENTER |

| ZOOM | [A] 5

Press **TRACE** and use the left and right arrow keys to move the trace cursor. Move the trace cursor to estimate the point where two lines intersect.

| ZOOM | [A] 3

Repeat the trace and zoom steps to get a more accurate estimate.

CASIO CFX-9850GA PLUS

From the main menu, choose GRAPH.

3 | EXE |

| (-) | 1 | EXE |

| (| 1 | ÷ | 4 |) | X,θ,T | + | 1 | EXE |

4 | X,θ,T | + | 4 | EXE |

| (-) | 4 | X,θ,T | + | 3 | EXE |

4 | X,θ,T | − | 8 | EXE |

| SHIFT | F3 | F3 | EXIT | F6 |

Press **SHIFT** **F1** and use the left and right arrow keys to move the trace cursor. Move the trace cursor to estimate the point where two lines intersect.

| SHIFT | F2 | F3 |

Repeat the trace and zoom steps to get a more accurate estimate.

NAME _____ DATE _____

Graphing Calculator Activity Keystrokes

For use with Technology Activity 4.6 on pages 248–249

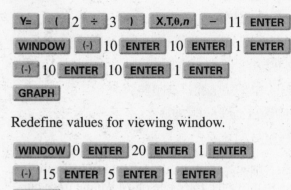

Keystrokes for Example 1

TI-82

| Y= | (| 2 | ÷ | 3 |) | X,T,θ | − | 11 | ENTER |

| WINDOW | (-) | 10 | ENTER | 10 | ENTER | 1 | ENTER |

| (-) | 10 | ENTER | 10 | ENTER | 1 | ENTER |

| GRAPH |

Redefine values for viewing window.

| WINDOW | 0 | ENTER | 20 | ENTER | 1 | ENTER |

| (-) | 15 | ENTER | 5 | ENTER | 1 | ENTER |

| GRAPH |

TI-83

| Y= | (| 2 | ÷ | 3 |) | X,T,θ,n | − | 11 | ENTER |

| WINDOW | (-) | 10 | ENTER | 10 | ENTER | 1 | ENTER |

| (-) | 10 | ENTER | 10 | ENTER | 1 | ENTER |

| GRAPH |

Redefine values for viewing window.

| WINDOW | 0 | ENTER | 20 | ENTER | 1 | ENTER |

| (-) | 15 | ENTER | 5 | ENTER | 1 | ENTER |

| GRAPH |

SHARP EL-9600c

| Y= | (| 2 | ÷ | 3 |) | X/θ/T/n | − | 11 |

| WINDOW | (-) | 10 | ENTER | 10 | ENTER | 1 | ENTER |

| (-) | 10 | ENTER | 10 | ENTER | 1 | ENTER |

| GRAPH |

Redefine values for viewing window.

| WINDOW | 0 | ENTER | 20 | ENTER | 1 | ENTER |

| (-) | 15 | ENTER | 5 | ENTER | 1 | ENTER |

| GRAPH |

CASIO CFX-9850GA PLUS

From the main menu, choose GRAPH.

| (| 2 | ÷ | 3 |) | X,θ,T | − | 11 | EXE |

| SHIFT | F3 | (-) | 10 | EXE | 10 | EXE | 1 | EXE |

| (-) | 10 | EXE | 10 | EXE | 1 | EXE | EXIT | F6 |

| GRAPH |

Redefine values for viewing window.

| SHIFT | F3 | 0 | EXE | 20 | EXE | 1 | EXE |

| (-) | 15 | EXE | 5 | EXE | 1 | EXE | EXIT | F6 |

Graphing Calculator Activity Keystrokes

For use with Technology Activity 4.6 on pages 248–249

Keystrokes for Example 2

TI-82

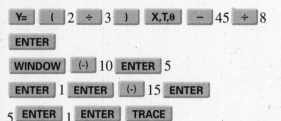

Use the left and right arrow keys to move the trace cursor. Move the trace cursor until the *x*-coordinate of the point is about −7. Use the zoom feature to get a more accurate estimate.

[ZOOM] 2 [ENTER]

Continue to use the trace and zoom features to get more accurate estimate.

TI-83

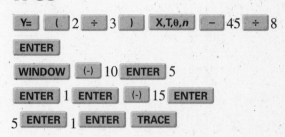

Use the left and right arrow keys to move the trace cursor. Move the trace cursor until the *x*-coordinate of the point is about −7. Use the zoom feature to get a more accurate estimate.

[ZOOM] 2 [ENTER]

Continue to use the trace and zoom features to get more accurate estimate.

SHARP EL-9600c

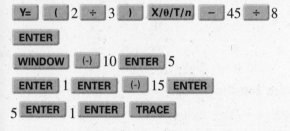

Use the left and right arrow keys to move the trace cursor. Move the trace cursor until the *x*-coordinate of the point is about −7. Use the zoom feature to get a more accurate estimate.

[ZOOM] [A]3

Continue to use the trace and zoom features to get more accurate estimate.

CASIO CFX-9850GA PLUS

From the main menu, choose GRAPH

[( ] 2 [÷] 3 [ )] [X,θ,T] [−] 45 [÷] 8 [EXE]

[(-)] 15 [EXE] 5 [EXE] 1 [EXE] [EXIT] [F6]

[SHIFT] [F1]

Use the left and right arrow keys to move the trace cursor. Move the trace cursor until the *x*-coordinate of the point is about −7. Use the zoom feature to get a more accurate estimate.

[F2] [F3]

Continue to use the trace and zoom features to get more accurate estimate.

NAME _____ DATE _____

Practice A
For use with pages 241–247

Find the slope and *y*-intercept of the line whose graph is shown.

1.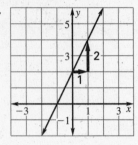

2.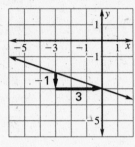

3.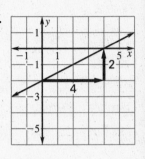

Find the slope and *y*-intercept of the graph of the equation.

4. $y = 7x + 3$

5. $y = 5x - 1$

6. $y = 7$

7. $y = -2x$

8. $y = \frac{1}{2}x + \frac{5}{2}$

9. $y = \frac{4x + 3}{2}$

Graph the equation. If necessary, write the equation in slope-intercept form first.

10. $y = x + 1$

11. $y = x - 6$

12. $y = 3x$

13. $y = -2x$

14. $y = 2x - 3$

15. $y = -5x - 2$

16. $y = 4$

17. $y = \frac{1}{2}x - 1$

18. $y = -\frac{2}{3}x + 2$

19. $y = \frac{3}{2}x + \frac{1}{2}$

20. $-3x + y = 8$

21. $x + y = 5$

Decide whether the graphs of the two equations are parallel lines.

22. $y = x + 3, y = x + 6$

23. $y = 2x - 3, y = -2x + 3$

24. $y = 4x - 1, y = 1 - 4x$

25. $3y = x - 12, 6y = 2x + 12$

Jogging   **Use the following information.**

Howard decides to start jogging every day at the track. The first week he jogs 4 laps. He adds 1 lap each week for 8 weeks. Let *l* represent the number of laps Howard runs and let *t* represent the time in weeks since he began jogging.

26. Make a table of values to record the number of laps Howard jogs from week 0, 1, 2, 3, . . . , 7.

27. Plot the ordered pairs. Draw a line through the points.

28. Find the slope. What does it represent?

Telephone Calls   **Use the following information.**

The cost of a long-distance telephone call is $.50 for the first minute and $.10 for each additional minute. Let *c* represent the total cost of a call that lasts *t* minutes.

29. Make a table of values to record the costs of calls that last 1, 2, 3, 4, 5, and 6 minutes.

30. Plot the ordered pairs. Draw a line through the points.

31. Find the slope. What does it represent?

Practice B

For use with pages 241–247

Find the slope and *y*-intercept of the graph of the equation.

1. $y = 7x + 1$

2. $y = -3x - 4$

3. $y = -4$

4. $y - 2x = 3.2$

5. $y = \dfrac{x + 3}{4}$

6. $2y = 6x + 16$

Graph the equation. If necessary, write the equation in slope-intercept form first.

7. $y = x + 5$

8. $y = 2x - 4$

9. $y = 3 - 2x$

10. $y = \frac{2}{3}x$

11. $y = \frac{1}{2}x - 4$

12. $y = -x - 3$

13. $y = -\frac{4}{5}x - \frac{1}{2}$

14. $y = \dfrac{x + 2}{3}$

15. $5x - 10y = -20$

16. $2y = 8$

17. $x + 10y - 3 = 7$

18. $2x + 4y = 6x - 6$

Decide whether the graphs of the two equations are parallel lines.

19. $y = 2x - 1, y = -2x + 1$

20. $y = 6x - 7, y = 3 + 6x$

21. $y = \dfrac{1}{4}x + 5, y = 4x - 7$

22. $y = -\dfrac{1}{2}x + \dfrac{3}{2}, y = \dfrac{8 - x}{2}$

23. $5x + y = -4, y - 5x = 6$

24. $7y = 2x + 7, 7y - 2x + 3 = 0$

Jogging   **Use the following information.**

Howard decides to start jogging every day at the track. The first week he jogs 6 laps. He adds 2 laps each week for 8 weeks. Let *l* represent the number of laps Howard runs and let *t* represent the time in weeks since he began jogging.

25. Plot points for the number of laps Howard jogs at one week intervals. Draw a line through the points.

26. Find the slope. What does it represent?

Telephone Calls   **Use the following information.**

The cost of a long-distance telephone call is $.85 for the first minute and $.05 for each additional minute. Let *c* represent the total cost of a call that lasts *t* minutes.

27. Plot points for the cost of calls in one minute intervals. Draw a line through the points.

28. Find the slope. What does it represent?

Weight Loss   **Use the following information.**

The graph at the right represents the weight loss of a wrestler as he prepares for the state meet.

29. Find the slope of the line. What does it represent?

30. Find the *w*-intercept. What does it represent?

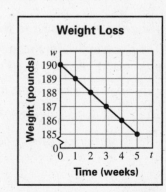

Weight Loss

Lesson 4.6

Practice C

For use with pages 241–247

Find the slope and *y*-intercept of the graph of the equation.

1. $y = -5x - 8$

2. $y = -\frac{3}{4}x$

3. $y + 6x = 1.8$

4. $4y - 2x = 3.2$

5. $y = \frac{3x - 7}{8}$

6. $9y = 5x - 15$

Graph the equation. If necessary, write the equation in slope-intercept form first.

7. $y = x - 3$

8. $y = 5x + 7$

9. $y = 9 - 3x$

10. $y = -\frac{4}{5}x$

11. $y = -\frac{2}{3}x + 4$

12. $y = -\frac{5}{4}x - \frac{3}{4}$

13. $y = -\frac{2x + 3}{9}$

14. $-3x + 6y = 18$

15. $-4x - 2y = -12$

16. $-x - y - 6 = 1$

17. $3(x + y) = 9$

18. $4(x + 2y - 1) = 8$

Decide whether the graphs of the two equations are parallel lines.

19. $y = x + 3, y = 6 + x$

20. $y = 3x - 2, y = \frac{1}{3}x + 4$

21. $2x + y = 3, 2x + y - 14 = 0$

22. $3y = x - 9, 3y + x - 3 = 0$

23. $3x + 2y = 1, -2y = 3x - 2$

24. $5y - x - 2 = 0, 2y - 10x + 2 = 0$

In Exercises 25–27, graph the situation described.

25. A full, 5 gallon container begins dripping at a rate of 0.5 gallons per hour. Let *t* represent the time in hours and *A* represent the amount remaining after *t* hours.

26. You start from home and drive 60 miles per hour for 3 hours. Let *t* represent the time in hours and *D* represent the distance from your home after *t* hours.

27. An airplane takes off and rises 800 feet per minute for 5 minutes. Let *t* represent the time in minutes and *h* represent the height of the plane after *t* minutes.

Telephone Calls **Use the following information.**

The cost of a long-distance telephone call is $.95 for the first minute and $.08 for each additional minute. Let *c* represent the total cost of a call that lasts *t* minutes.

28. Plot points for the cost of calls in one minute intervals. Draw a line through the points.

29. Find the slope. What does it represent?

Weight Loss **Use the following information.**

The graph at the right represents the weight loss of a wrestler as he prepares for the state meet.

30. Find the slope of the line. What does it represent?

31. Find the *w*-intercept. What does it represent?

32. If the wrestler lost 2 pounds per week, how would the graph change?

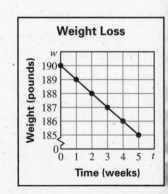

Weight Loss

NAME _____ DATE _____

Reteaching with Practice

For use with pages 241–247

GOAL Graph a linear equation in slope-intercept form and interpret equations in slope-intercept form.

VOCABULARY

The linear equation $y = mx + b$ is written in **slope-intercept form.** The slope of the line is m. The y-intercept is b.

Two different lines in the same plane are **parallel** if they do not intersect. Any two nonvertical lines are parallel if and only if they have the same slope (all vertical lines are parallel).

EXAMPLE 1 *Writing Equations in Slope-Intercept Form*

EQUATION	SLOPE-INTERCEPT FORM	SLOPE	y-INTERCEPT
a. $y = 3x$	$y = 3x + 0$	$m = 3$	$b = 0$
b. $y = \dfrac{2x - 3}{5}$	$y = \dfrac{2}{5}x - \dfrac{3}{5}$	$m = \dfrac{2}{5}$	$b = -\dfrac{3}{5}$
c. $4x + 8y = 24$	$y = -0.5x + 3$	$m = -0.5$	$b = 3$

Exercises for Example 1

Write the equation in slope-intercept form. Find the slope and the *y*-intercept

1. $y = -3x$ **2.** $x + y - 5 = 0$ **3.** $3x + y = 5$

4. $y = \dfrac{-x + 7}{3}$ **5.** $y = 2$ **6.** $x + 4y - 4 = 0$

7. Which two lines in Exercises 1–6 are parallel? Explain.

EXAMPLE 2 *Graphing Using Slope and y-Intercept*

Graph the equation $5x - y = 3$.

SOLUTION

Write the equation in slope-intercept form: $y = 5x - 3$

Find the slope and the y-intercept: $m = 5$ and $b = -3$.

Plot the point $(0, b)$. Draw a slope triangle to locate a second point on the line.

$$m = \frac{5}{1} = \frac{\text{rise}}{\text{run}}$$

Draw a line through the two points.

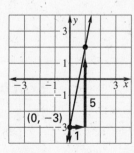

Reteaching with Practice

For use with pages 241–247

Exercises for Example 2

Write the equation in slope-intercept form. Then graph the equation.

8. $6x - y = 0$
9. $x + 3y - 3 = 0$
10. $5x + y = 4$

11. $x + 3y - 6 = 0$
12. $2x + y - 9 = 0$
13. $x + 2y + 8 = 0$

EXAMPLE 3 *Using Slope-Intercept Form to Solve a Real-Life Problem*

During the summer you work for a lawn care service. You are paid $5 per day, plus an hourly rate of $1.50.

a. Using w to represent daily wages and h to represent the number of hours worked daily, write an equation that models your total wages for one day's work.

b. Find the slope and the y-intercept of the equation.

c. What does the slope represent?

d. Graph the equation, using the slope and the y-intercept.

SOLUTION

a. Using w to represent daily wages and h to represent the number of hours worked daily, the equation that models your total wages for one day's work is $w = 1.50h + 5$.

b. The slope of the equation is 1.50 and the y-intercept is 5.

c. The slope represents the hourly rate.

d.

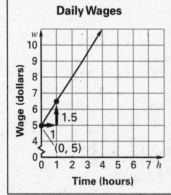

Exercises for Example 3

14. Rework Example 3 if you are paid $4 per day, plus an hourly rate of $1.75.

15. Rework Example 3 if you are paid $6 per day, plus an hourly rate of $1.25.

NAME _____ DATE _____

Quick Catch-Up for Absent Students

For use with pages 240–249

The items checked below were covered in class on (date missed) _____

Activity 4.6: Graphing Families of Linear Equations (p. 240)

_____ **Goal:** Explore the relationships that exist between members of families of equations.

Lesson 4.6: Quick Graphs Using Slope-Intercept Form

_____ **Goal 1:** Graph a linear equation in slope-intercept form. (pp. 241–242)

Material Covered:

_____ Activity: Investigating Slope-Intercept Form

_____ Example 1: Writing Equations in Slope-Intercept Form

_____ Example 2: Graphing Using Slope and *y*-Intercept

_____ Example 3: Identifying a Family of Parallel Lines

Vocabulary:

slope-intercept form, p. 241   parallel lines, p. 242

_____ **Goal 2:** Graph and interpret equations in slope-intercept form in a real-life problem. (p. 243)

Material Covered:

_____ Example 4: Graphing Using Slope-Intercept Form

Activity 4.6: Graphing a Linear Equation (pp. 248–249)

_____ **Goal:** Graph a linear equation and find solutions using a graphing calculator or a computer.

_____ Student Help: Keystroke Help

_____ Student Help: Study Tip

_____ Other (specify) _____

_____

Homework and Additional Learning Support

_____ Textbook (specify) pp. 244–247 _____

_____

_____ *Reteaching with Practice* worksheet (specify exercises)_____

_____ *Personal Student Tutor* for Lesson 4.6

Interdisciplinary Application

For use with pages 241–247

Mount Everest

GEOGRAPHY   Mount Everest, the highest mountain in the world, is located 160 kilometers northeast of Katmandu, Nepal, in the Himalayan mountain range between India and Tibet. The Tibetan name for Mount Everest is Chomolungma, which means "goddess mother of the world." The name Everest, given in 1865, honors Sir George Everest, the British surveyor who established the location and approximate height of the mountain—at that time estimated to be 29,002 feet.

Base camps on the Nepalese side of the mountain are at around 18,000 feet, an altitude of about 2,300 feet less than that of Mount McKinley, the highest mountain in North America. After the opening of Tibet to outsiders in 1920, a number of attempts were made to climb Mount Everest. The year-round snow, extreme cold, high winds, and low oxygen supply made the climb extremely dangerous. In 1953, however, Edmund Hillary of New Zealand and Tenzing Norgay, a Nepalese Sherpa tribesman, succeeded in reaching the peak. Through their efforts, the official height of Mount Everest was given as 29,028 feet. In 1999, new calculations using the Global Positioning System (GPS) put the height officially at 29,035 feet.

Sherpas are Tibetan Buddhists. Sherpa climbers are extremely brave and self-less in the face of danger. In 1991, the Sherpas, who had been acting as guides for trekkers from other countries for years, organized an all-Sherpa expedition to the top to honor all Sherpas. When they reached the summit, they radioed down to the village below, "We're on top!" Out of the total number of climbers who have reached the peak of Mount Everest, about one fourth have been Sherpas. Two of them, Ang Rita and Appa Sherpa, have stood on the peak at least 10 times each.

1. The peak of Mount Everest is approximately 6.25 miles along the horizontal from a base camp on the Nepalese side of the mountain. Using 29,035 feet as an estimation of the altitude of Mount Everest's peak and 18,000 feet as the altitude for the base camp, what is the slope of the climb from the base camp to the peak? Assume the base camp is located on the y-axis. (*Hint:* One mile is equal to 5280 feet.)

2. Using the information from Exercise 1, write the equation of the line that models their climb in slope-intercept form. Write in words what the slope and the y-intercept mean in the equation.

3. Graph the line using the slope and y-intercept from Exercise 2.

4. A group of climbers begin their journey to Mount Everest's peak from the base camp discussed in Exercise 1. At what elevation will they be after 4 miles (in the horizontal plane) have been covered? Round your answer to the nearest foot.

NAME _____ DATE _____

Challenge: Skills and Applications

1. Solve the equation $Ax + By = C$ for y.

2. Based on the result from Exercise 1, what is the slope of the graph of $Ax + By = C$? What is the y-intercept?

Use the result from Exercise 2 to find the slope (m) and the y-intercept (b) of the graph of the equation, without rewriting the equation in slope-intercept form.

3. $5x - 3y = 15$

4. $-2x + 7y = 14$

5. $9x + 6y = -18$

6. $4x + 5y = 7$

7. In order to graph the equation $3x - 4y = 12$ using the intercepts method taught in Lesson 4.3, what two points do you plot?

8. In order to graph the equation $3x - 4y = 12$ using the slope-intercept method, what two points do you plot?

9. Complete the table and look for a pattern to determine when you plot the same two points with the intercepts method and the slope-intercept method.

Equation	Points plotted	
	With intercepts method	*With slope-intercept method*
$2x + 5y = 10$		
$4x + 5y = 10$		
$7x + 2y = -14$		
$4x + 6y = -12$		
$-9x + 4y = -36$		

10. What relationship do A, B, and C seem to have in equations $Ax + By = C$ for which you plot the same two points when graphing with the intercepts method or with the slope-intercept method?

NAME _____ DATE _____

Quiz 2

For use after Lessons 4.4–4.6

1. Find the slope of the line passing through the points $(4, -4)$ and $(6, -5)$. *(Lesson 4.4)*

2. Find the value of y so that the line passing through the points $(0, y)$ and $(3, 6)$ has a slope of $-\frac{1}{3}$. *(Lesson 4.4)*

3. Complete the sentence. The slope of a vertical line is _____. *(Lesson 4.4)*

4. The variables x and y vary directly. When $x = 12$, $y = 9$. Write an equation that relates the variables. *(Lesson 4.5)*

5. Graph the equation $y = -\frac{1}{2}x$. Find the constant of variation and the slope of the direct variation model. *(Lesson 4.5)*

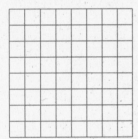

For Exercise 5

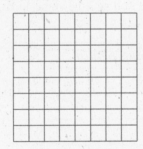

For Exercise 6

6. Write the equation $-\frac{1}{2}x + \frac{1}{2}y = 5$ in slope-intercept form. Then graph the equation. *(Lesson 4.6)*

7. Decide whether the graphs of $3x - 6y = 12$ and $y = \dfrac{x + 4}{2}$ are parallel lines. Explain your answer. *(Lesson 4.6)*

Answers

1. _____

2. _____

3. _____

4. _____

5. Use grid at left. _____

_____

_____

6. Use grid at left. _____

_____

7. _____

_____

_____

TEACHER'S NAME _____ CLASS _____ ROOM _____ DATE _____

Lesson Plan

1-day lesson (See *Pacing the Chapter,* TE pages 200C–200D)   **For use with pages 250–255**

GOALS 1. Solve a linear equation graphically.
2. Use a graph to approximate a solution to a real-life problem.

State/Local Objectives _____

_____

✓ Check the items you wish to use for this lesson.

STARTING OPTIONS
____ Homework Check: TE page 244; Answer Transparencies
____ Warm-Up or Daily Homework Quiz: TE pages 250 and 247, CRB page 97, or Transparencies

TEACHING OPTIONS
____ Motivating the Lesson: TE page 251
____ Lesson Opener (Graphing Calculator): CRB page 98 or Transparencies
____ Graphing Calculator Activity with Keystrokes: CRB page 99
____ Examples 1–5: SE pages 250–252
____ Extra Examples: TE pages 251–252 or Transparencies; Internet
____ Closure Question: TE page 252
____ Guided Practice Exercises: SE page 253

APPLY/HOMEWORK
Homework Assignment
____ Basic   11–13, 14–36 even, 44, 45, 50, 55, 60, 65, 66
____ Average   11–13, 14–36 even, 50, 55, 60, 65, 66
____ Advanced   11–13, 14–42 even, 46–51, 55, 60, 65, 66

Reteaching the Lesson
____ Practice Masters: CRB pages 100-102 (Level A, Level B, Level C)
____ Reteaching with Practice: CRB pages 103–104 or Practice Workbook with Examples
____ Personal Student Tutor

Extending the Lesson
____ Applications (Real-Life): CRB page 106
____ Challenge: SE page 255: CRB page 107 or Internet

ASSESSMENT OPTIONS
____ Checkpoint Exercises: TE pages 251–252 or Transparencies
____ Daily Homework Quiz (4.7): TE page 255, CRB page 110, or Transparencies
____ Standardized Test Practice: SE page 255; TE page 255; STP Workbook; Transparencies

Notes _____

_____

_____

_____

Lesson 4.7

TEACHER'S NAME _____ CLASS _____ ROOM _____ DATE _____

Lesson Plan for Block Scheduling

Half-day lesson (See *Pacing the Chapter,* TE pages 200C–200D)   For use with pages 250–255

GOALS 1. **Solve a linear equation graphically.**
2. **Use a graph to approximate a solution to a real-life problem.**

State/Local Objectives _____

_____

_____

✓ **Check the items you wish to use for this lesson.**

STARTING OPTIONS

____ Homework Check: TE page 244; Answer Transparencies
____ Warm-Up or Daily Homework Quiz: TE pages 250 and
          247, CRB page 97, or Transparencies

CHAPTER PACING GUIDE	
Day	**Lesson**
1	Assess Ch. 3; 4.1 (all)
2	4.2 (all)
3	4.3 (all)
4	4.4 (all)
5	4.5 (all); 4.6 (begin)
6	4.6 (end); **4.7 (all)**
7	4.8 (all)
8	Review/Assess Ch. 4

TEACHING OPTIONS

____ Motivating the Lesson: TE page 251
____ Lesson Opener (Graphing Calculator): CRB page 98 or Transparencies
____ Graphing Calculator Activity with Keystrokes: CRB page 99
____ Examples 1–5: SE pages 250–252
____ Extra Examples: TE pages 251–252 or Transparencies; Internet
____ Closure Question: TE page 252
____ Guided Practice Exercises: SE page 253

APPLY/HOMEWORK

Homework Assignment (See also assignment for Lesson 4.6.)

____ Block Schedule: 11–13, 14–36 even, 44–48, 50, 55, 60, 65, 66

Reteaching the Lesson

____ Practice Masters: CRB pages 100–102 (Level A, Level B, Level C)
____ Reteaching with Practice: CRB pages 103–104 or Practice Workbook with Examples
____ Personal Student Tutor

Extending the Lesson

____ Applications (Real-Life): CRB page 106
____ Challenge: SE page 255: CRB page 107 or Internet

ASSESSMENT OPTIONS

____ Checkpoint Exercises: TE pages 251–252 or Transparencies
____ Daily Homework Quiz (4.7): TE page 255, CRB page 110, or Transparencies
____ Standardized Test Practice: SE page 255; TE page 255; STP Workbook; Transparencies

Notes _____

_____

_____

NAME _____ DATE _____

WARM-UP EXERCISES

For use before Lesson 4.7, pages 250–255

Solve each equation.

1. $-4x + 1 = 17$ **2.** $5 - 3y = 8$

What is the *x*-intercept of each function?

3. $4x - 8 = y$ **4.** $y = -x - 3$ **5.** $x + y = -5$

....

DAILY HOMEWORK QUIZ

For use after Lesson 4.6, pages 240–249

1. Find the slope and the *y*-intercept of $4x - 2y = 8$.

2. Write $x + y + 3 = 0$ in slope-intercept form. Then graph
the equation.

3. Are the graphs of the equations given below parallel?
Explain your answer.
$y - 2x = 5$ and $2y = 4x$

4. Describe the figure formed by the intersection of the graphs of
the equations. What is its area?
$x = -3, x = 3, y = -3, y = 3$

NAME _____ DATE _____

Graphing Calculator Lesson Opener

For use with pages 250–255

You know how to solve equations algebraically. You can also use a graph to solve a linear equation. For example, consider the equation $3x + 5 = -31$.

1. Write the equation in the form $ax + b = 0$.

2. Write the function $y = ax + b$, which is called the *related function*.

3. Use a graphing calculator to graph the function in Step 2. Adjust the viewing window as necessary until you can see the x- and y-intercepts clearly.

4. Find the x-intercept of the line. Substitute this value into the original equation. What do you notice?

5. Solve the equation algebraically. Compare your solution to the x-intercept you found in Step 4. What conclusion can you make?

6. Repeat Steps 1–5 for the following equations. Be sure to use parentheses around fractions when entering the equations into your calculator.

    a. $10 - \dfrac{1}{2}x = 15$ **b.** $\dfrac{1}{4}x + \dfrac{3}{4} = 2$

7. Use the results of your explorations to make a conclusion about using a graph to solve a linear equation.

Graphing Calculator Activity Keystrokes

For use with pages 251 and 254

Keystrokes for Example 3

TI-82

`Y=` 2.65 `(` 4 `X,T,θ` `–` 9 `)` `+`
8.85 `–` 7.6 `X,T,θ`
`ZOOM` 6 `TRACE`

TI-83

`Y=` 2.65 `(` 4 `X,T,θ,n` `–` 9 `)` `+`
8.85 `–` 7.6 `X,T,θ,n`
`ZOOM` 6 `TRACE`

SHARP EL-9600c

`Y=` 2.65 `(` 4 `X/θ/T/n` `–` 9 `)` `+`
8.85 `–` 7.6 `X/θ/T/n`
`ZOOM` [A]5 `TRACE`

CASIO CFX-9850GA PLUS

From the main menu, choose GRAPH
2.65 `(` 4 `X,θ,T` `–` 9 `)` `+`
8.85 `–` 7.6 `X,T,θ` `EXE` `SHIFT` `F3` `F3` `EXIT`
`F6` `F1`

Keystrokes for Exercise 38

TI-82

`Y=` `X,T,θ` `–` 7 `ENTER`
`ZOOM` 6 `TRACE`

TI-83

`Y=` `X,T,θ,n` `–` 7 `ENTER`
`ZOOM` 6 `TRACE`

SHARP EL-9600c

`Y=` `X/θ/T/n` `–` 7 `ENTER`
`ZOOM` [A]5 `TRACE`

CASIO CFX-9850GA PLUS

From the main menu, choose GRAPH
`X,θ,T` `–` 7 `EXE` `SHIFT` `F3` `F3`
`EXIT` `F6` `F1`

Practice A

For use with pages 250–255

Identify the x-intercept of the line whose graph is shown.

1.
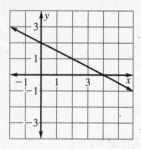

2.
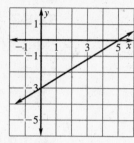

3.
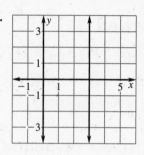

In Exercises 4–9, match the one-variable equation with its related function.

A. $3x = 6$

B. $3x - 4 = 6$

C. $5x + 2 = 2x$

D. $3x = 5x - 6$

E. $2x + 2 = 6$

F. $-x = x + 8$

4. $y = 3x - 10$

5. $y = 2x - 6$

6. $y = -3x + 6$

7. $y = 2x - 4$

8. $y = 2x + 8$

9. $y = 3x + 2$

Write the equation in the form $ax + b = 0$. Then write the related function $y = ax + b$.

10. $5x + 2 = 7$

11. $3 - 3x = 9$

12. $7 + 4x = 15$

13. $8x - 9 = 4x$

14. $-6x + 1 = 5x - 4$

15. $7 - 3x = 10 + 5x$

Solve the equation algebraically. Check your solution graphically.

16. $5x + 3 = -2$

17. $-2x + 13 = 7$

18. $-x = -4$

19. $3x - 5 = 13$

20. $\frac{1}{3}x + 4 = 9$

21. $-\frac{1}{2}x - 8 = 12$

22. $5x - 2 = -6$

23. $-3x + 5 = -4x + 8$

24. $4x - \frac{1}{2} = 7\frac{1}{2}$

Solve the equation graphically. Check your solution algebraically.

25. $-2x = 6$

26. $3x + 5 = 4$

27. $4 + 5x = 9$

28. $7 - 5x = -2$

29. $5x + 6 = 2x + 2$

30. $3x + 4 = -2 - 5x$

31. *High School Alumni* The number of students A who graduated from Monroe High School between 1990 and 2000 can be modeled by the equation $A = 250t + 2000$, where t is the number of years since 1990. In what year were there 3250 alumni? Solve algebraically and graphically.

32. *Cassette Sales* The number of cassette tapes y sold at a local music store between 1985 and 2000 can be modeled by the equation $y = -350t + 6000$, where t is the number of years since 1985. In what year did the store sell 1100 cassettes? Solve algebraically and graphically.

Lesson 4.7

NAME _____ DATE _____

Practice B

For use with pages 250–255

In Exercises 1–6, match the one-variable equation with its related function.

A. $-3x = 2$ **B.** $-3x - 6 = -2$ **C.** $-5x = -x + 3$

D. $3 = 4x$ **E.** $3x + 3 = 5$ **F.** $2x = -x + 4$

1. $y = 3x - 4$ **2.** $y = -4x - 3$ **3.** $y = -3x - 4$

4. $y = -4x + 3$ **5.** $y = 3x - 2$ **6.** $y = -3x - 2$

Write the equation in the form $ax + b = 0$. Then write the related function $y = ax + b$.

7. $-4 = -5x$ **8.** $2 - 3x = 9$ **9.** $-7 + 4x = 13$

10. $8x - 9 = 2x$ **11.** $-6x + 1 = 10x - 4$ **12.** $6 - \frac{1}{4}x = \frac{3}{4}x - 6$

Solve the equation algebraically. Check your solution graphically.

13. $5x - 14 = -4$ **14.** $-3x + 13 = 1$ **15.** $-x = 3$

16. $6x + 2 = -4x$ **17.** $-12 - 7x = 3x + 8$ **18.** $-\frac{1}{2}x - 9 = -7$

19. $\frac{2}{5}x - 2 = -6$ **20.** $-\frac{3}{4}x + \frac{4}{5} = \frac{1}{4}x - \frac{1}{5}$ **21.** $\frac{1}{3} = 5 - \frac{2}{3}x$

Solve the equation graphically. Check your solution algebraically.

22. $2x + 6 = 4$ **23.** $3 - 8x = -13$ **24.** $4x - 2 = -4x$

25. $3x + 15 = -9 + 7x$ **26.** $-5x + 7 = 2x - 14$ **27.** $\frac{1}{3}x + 2 = -2$

28. Fundraiser Your school's math club is having a car wash to raise $600 for a trip. The amount A that the club raised can be modeled by the equation $A = 4.5n - 120$, where n is the number of cars washed. How many cars must the club wash to be able to go on the trip?

29. Production Costs Joan has a small business printing promotional brochures in her home. Her monthly cost of producing x brochures can be modeled by the function $y = 0.2x + 120$. In July, her cost was $288. How many brochures did she print that month? Solve algebraically and graphically.

30. Geometry The line $y = -x + 25$ represents all possible dimensions of a rectangle with a perimeter of 50 centimeters, where x represents the length and y represents the width. What is the length of a rectangle that has a width of 15 centimeters? Solve algebraically and graphically.

31. Motion Pictures The amount, A (in billions of dollars), spent to see motion pictures from 1990 through 1995 can be modeled by the equation $A = 3.7t + 39.27$, where $t = 0$ represents 1990. According to this model, in what year will the amount spent to see motion pictures reach $80 billion?

Algebra 1
Chapter 4 Resource Book

Practice C

For use with pages 250–255

Write the equation in the form $ax + b = 0$. Then write the related function $y = ax + b$.

1. $5x + 2 = 8$ **2.** $5 - 3x = -11$ **3.** $7 + 4x = 15 - 2x$

4. $8x - 9 = 4.6x$ **5.** $-6x + 1 = 5.5x - 4$ **6.** $7 - \frac{2}{5}x = 10 + \frac{3}{5}x$

Solve the equation algebraically. Check your solution graphically.

7. $6x + 3 = -9$ **8.** $-2x - 16 = -7$ **9.** $8 - 3x = 5x$

10. $3x - 15 = 13 - x$ **11.** $\frac{2}{3}x + 4 = -10$ **12.** $-\frac{1}{2}x - 5 = \frac{3}{4}x$

13. $x - 6 = \frac{2}{5}x$ **14.** $-\frac{3}{4}x + \frac{5}{6} = \frac{1}{4}x - \frac{1}{6}$ **15.** $\frac{3}{2}x - \frac{7}{2} = \frac{1}{2}$

Solve the equation graphically. Check your solution algebraically.

16. $3x - 9 = -12$ **17.** $5 + 4x = 2x$ **18.** $\frac{1}{2}x - 6 = 2$

19. $-\frac{1}{3}x = \frac{2}{3}x - 1$ **20.** $-x + 7 = 11 - 5x$ **21.** $\frac{1}{6}x - 3 = 3 + \frac{5}{6}x$

Use a graphing calculator to find the solution of the equation.

22. $-2(x + 5) = 6(x + 2)$ **23.** $6(x + 3) = 2(x + 5)$ **24.** $6(x + 2) = 5(x + 2)$

25. $\frac{6}{5}(x + 2) = 5x + \frac{1}{2}$ **26.** $x - \frac{1}{3} = \frac{5}{4}(x - 6)$ **27.** $\frac{3}{2}(x - 10) = \frac{2}{3}(x - 12)$

28. *Production Costs* Joan has a small business printing promotional brochures in her home. Her monthly cost of producing x brochures can be modeled by the function $y = 0.25x + 137.5$. In July, her cost was $375. How many brochures did she print that month? Solve algebraically and graphically.

29. *Geometry* The line $y = -\frac{1}{2}x + \frac{13}{2}$ represents all possible dimensions of an isosceles triangle with a perimeter of 13 inches, where x represents the base and y represents the two equal sides. What is the length of the base of an isosceles triangle with sides of 5 inches each? Solve algebraically and graphically.

30. *U.S. Population* The population, P (in millions), of the United States from 1990 through 1996 can be modeled by the equation $P = 2.74t + 249.23$, where $t = 0$ represents 1990. According to this model, in what year will the population of the United States reach 300 million?

Books and Maps **Use the following information.**

The amount, A (in billions of dollars), spent on books and maps from 1990 through 1995 can be modeled by the equation $A = 0.94t + 16.17$, where $t = 0$ represents 1990.

31. In what year was $19 billion spent on books and maps?

32. According to this model, in what year will the amount spent on books and maps reach $35 billion?

Lesson 4.7

NAME _____ DATE _____

Reteaching with Practice

For use with pages 250–255

GOAL Solve a linear equation graphically and use a graph to approximate solutions in real-life problems.

The first step in solving a linear equation graphically is to write the equation in the form $ax + b = 0$. Next, write the related function $y = ax + b$. Finally, graph the equation $y = ax + b$. The solution of $ax + b = 0$ is the x-intercept of $y = ax + b$.

EXAMPLE 1 *Solving an Equation Graphically*

Solve $3x - 1 = 5$ graphically. Check your solution algebraically.

SOLUTION

1 Write the equation in the form $ax + b = 0$.

| $3x - 1 = 5$ | Write original equation. |
| $3x - 6 = 0$ | Subtract 5 from each side. |

2 Write the related function $y = 3x - 6$.

3 Graph the equation $y = 3x - 6$.
The x-intercept appears to be 2.

4 Use substitution to check your solution.

$$3x - 1 = 5$$
$$3(2) - 1 \overset{?}{=} 5$$
$$6 - 1 = 5 \qquad \leftarrow \text{True statement}$$

The solution of $3x - 1 = 5$ is 2.

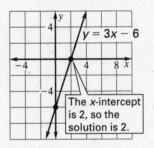

The x-intercept is 2, so the solution is 2.

Exercises for Example 1

Solve the equation graphically. Check your solution algebraically.

1. $5x + 2 = 7$ **2.** $-3x = 15$ **3.** $2 - x = 5$

4. $8 + 2x = -2x$ **5.** $0.5x + 1 = 3$ **6.** $3x + 6 = 11 - 2x$

Lesson 4.7

NAME _____ DATE _____

Reteaching with Practice

For use with pages 250–255

EXAMPLE 2 *Approximating a Real-Life Solution*

Based on data from 1989 to 1995, a model for the number n (in millions) of women in the civilian labor force in the United States is $n = 0.821t + 55.7$, where t is the number of years since 1989. According to this model, in what year will the United States have 72 million women in the civilian labor force? *(Source: Department of Labor, Bureau of Labor Statistics)*

SOLUTION

Substitute 72 for n in the linear model. Solve the resulting linear equation, $72 = 0.821t + 55.7$, to answer the question.

Write the equation in the form $ax + b = 0$.

$$72 = 0.821t + 55.7$$

$$0 = 0.821t - 16.3$$

Graph the related function $n = 0.821t - 16.3$.

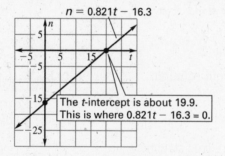

The t-intercept is about 19.9. Because t is the number of years since 1989, you can estimate that there will be 72 million women in the civilian labor force about 20 years after 1989, or about 2009.

Exercise for Example 2

7. Based on data from 1992 to 1995, a model for the United States consumer price index n is $n = 4t + 140.35$, where t is the number of years since 1992. According to this model, in what year will the United States have a consumer price index of 180.4?

Lesson 4.7

NAME _____ DATE _____

Quick Catch-Up for Absent Students

For use with pages 250–255

The items checked below were covered in class on (date missed) _____

Lesson 4.7: Solving Linear Equations Using Graphs

____ **Goal 1:** Solve a linear equation graphically. (pp. 250–251)

Material Covered:

____ Example 1: Using a Graphical Check for a Solution

____ Student Help: Study Tip

____ Example 2: Solving an Equation Graphically

____ Example 3: Using a Graphing Calculator

____ **Goal 2:** Use a graph to approximate a solution to a real-life problem. (p. 252)

Material Covered:

____ Example 4: Approximating a Real-Life Solution

____ Example 5: Solving by Graphing Both Sides

____ Other (specify) _____

_____

Homework and Additional Learning Support

____ Textbook (specify) pp. 253–255 _____

_____

____ Internet: Extra Examples at www.mcdougallittel.com

____ *Reteaching with Practice* worksheet (specify exercises) _____

____ *Personal Student Tutor* for Lesson 4.7

NAME _____ DATE _____

Real-Life Applications:
When Will I Ever Use This?

For use with pages 250–255

Women's NCAA Basketball

The first recorded women's collegiate basketball game was in 1896 between Stanford and the University of California at Berkeley. The International Women's Sports Federation began including basketball in its version of the Olympics in 1924. In 1926, the Amateur Athletic Union began sponsoring an annual women's tournament. The World Amateur Basketball Championships began in 1953 with the United States taking the first gold medal. Women's basketball finally debuted at the Olympics in 1976 in Montreal. A tremendous increase in the number of U.S. women participating in basketball (and other sports) occurred in 1972 when Title IX of the Education Amendments was enacted in 1972. One of the requirements of Title IX is equal opportunities for men and women in athletics at school and collegiate levels. After the Association for Intercollegiate Athletics for Women became part of the National Collegiate Athletic Association (NCAA) in the early 1980s, the attendance at women's basketball games increased through the years and had quadrupled by 1996.

1. Based on data from 1990 to 1996, a model for the number n (in thousands) of spectators at women's NCAA basketball games each year is $n = 432.07t + 2746.43$, where t is the number of years since 1990. According to this model, in what year will women's basketball have 8 million spectators?

2. Did you solve Exercise 1 algebraically or graphically? Give a reason for your choice.

3. Graph the model from Exercise 1. Use the graph to estimate the attendance in the year 2005.

4. In 1996, 28 million people attended men's NCAA basketball games. Based on the model from Example 1, predict when the attendance at NCAA women's basketball games will be 28 million.

5. Participation by girls in high school sports has grown since Title IX was enacted in 1972. Based on data from 1990 to 1997, a model for the number n (in thousands) of girls participating in high school sports each year is $n = 91.71t + 1791.52$, where t is the number of years since 1990. According to this model, in what year will 2800 thousand girls be participating in high school sports?

6. Graph the model from Exercise 5. Use the graph to estimate the number of girls participating in high school sports in the year 2004.

Algebra 1
Chapter 4 Resource Book

Challenge: Skills and Applications

For use with pages 250–255

In Exercises 1–8, use the following information.

In 1992, 7.6 million music videos were shipped, and in 1996, 16.9 million were shipped.

1. Find the rate of change in millions of music videos shipped per year.

2. Write an equation to model the millions of music videos *y* shipped *x* years after 1992.

3. According to the model from Exercise 2, how many music videos were shipped in 1994?

4. The actual number of music videos shipped in 1994 was 11.2 million. How close did the model predict this number?

5. Write an equation you can solve to find the year that the model predicts 25 million music videos will be shipped.

6. Solve the equation from Exercise 5 graphically. Interpret the result.

7. Write an equation you can solve to find the year that the model predicts 40 million music videos will be shipped.

8. Solve the equation from Exercise 7 graphically. Interpret the result.

In Exercises 9–12, use the following information.

In 1992, 407.5 million CD albums were shipped, and in 1996, 778.9 million were shipped.

9. Find the rate of change in millions of CD albums shipped per year.

10. Write an equation to model the millions of CD albums *y* shipped *x* years after 1992.

11. Write an equation you can solve to find the year that the model predicts one billion CD albums will be shipped.

12. Solve the equation from Exercise 11 graphically. Interpret the result.

Teacher's Name _____ Class _____ Room _____ Date _____

Lesson Plan

2-day lesson (See *Pacing the Chapter,* TE pages 200C–200D)   **For use with pages 256–262**

GOALS   **1. Identify when a relation is a function.**
2. Use function notation to represent real-life situations.

State/Local Objectives _____

_____

✓ **Check the items you wish to use for this lesson.**

STARTING OPTIONS
_____ Homework Check: TE page 253; Answer Transparencies
_____ Warm-Up or Daily Homework Quiz: TE pages 256 and 255, CRB page 110, or Transparencies

TEACHING OPTIONS
_____ Motivating the Lesson: TE page 257
_____ Lesson Opener (Application): CRB page 111 or Transparencies
_____ Examples: Day 1: 1–3, SE pages 256–257; Day 2: 4, SE page 258
_____ Extra Examples: Day 1: TE page 257 or Transp.; Day 2: TE page 258 or Transp.
_____ Closure Question: TE page 258
_____ Guided Practice: SE page 259 Day 1: Exs. 1–9; Day 2: Ex. 10

APPLY/HOMEWORK
Homework Assignment
_____ Basic   Day 1: 11–19, 20–28 even, 29–31, 32–48 even; Day 2: 33–49 odd, 50–52, 56–58, 65, 70,
         72–79, Quiz 3: 1–19
_____ Average   Day 1: 11–19, 20–28 even, 29–31, 32–48 even; Day 2: 33–49 odd, 50–53, 56–58, 65,
         70, 72–79, Quiz 3: 1–19
_____ Advanced   Day 1: 11–19, 20–28 even, 29–31, 32–48 even; Day 2: 33–49 odd, 53–55, 59–62, 65,
         70, 72–79, Quiz 3: 1–19

Reteaching the Lesson
_____ Practice Masters: CRB pages 112–114 (Level A, Level B, Level C)
_____ Reteaching with Practice: CRB pages 115–116 or Practice Workbook with Examples
_____ Personal Student Tutor

Extending the Lesson
_____ Applications (Interdisciplinary): CRB page 118
_____ Challenge: SE page 261: CRB page 119 or Internet

ASSESSMENT OPTIONS
_____ Checkpoint Exercises: Day 1: TE page 257 or Transp.; Day 2: TE page 258 or Transp.
_____ Daily Homework Quiz (4.8): TE page 262, or Transparencies
_____ Standardized Test Practice: SE page 261; TE page 262; STP Workbook; Transparencies
_____ Quiz (4.7–4.8): SE page 262

Notes _____

_____

_____

TEACHER'S NAME _____ CLASS _____ ROOM _____ DATE _____

Lesson Plan for Block Scheduling

1-day lesson (See *Pacing the Chapter,* TE pages 200C–200D)   **For use with pages 256–262**

GOALS   1. **Identify when a relation is a function.**
2. **Use function notation to represent real-life situations.**

State/Local Objectives _____

_____

_____

✓ **Check the items you wish to use for this lesson.**

STARTING OPTIONS
____ Homework Check: TE page 253; Answer Transparencies
____ Warm-Up or Daily Homework Quiz: TE pages 256 and
          255, CRB page 110, or Transparencies

TEACHING OPTIONS
____ Motivating the Lesson: TE page 257
____ Lesson Opener (Application): CRB page 111 or Transparencies
____ Examples 1–4: SE pages 256–258
____ Extra Examples: TE pages 257–258 or Transparencies
____ Closure Question: TE page 258
____ Guided Practice Exercises: SE page 259

APPLY/HOMEWORK
Homework Assignment
____ Block Schedule: 11–19, 20–28 even, 29–31, 32–49, 50–53, 56–58, 65, 70, 72–79, Quiz 3: 1–19

Reteaching the Lesson
____ Practice Masters: CRB pages 112–114 (Level A, Level B, Level C)
____ Reteaching with Practice: CRB pages 115–116 or Practice Workbook with Examples
____ Personal Student Tutor

Extending the Lesson
____ Applications (Interdisciplinary): CRB page 118
____ Challenge: SE page 261: CRB page 119 or Internet

ASSESSMENT OPTIONS
____ Checkpoint Exercises: TE pages 257–258 or Transparencies
____ Daily Homework Quiz (4.8): TE page 262 or Transparencies
____ Standardized Test Practice: SE page 261; TE page 262; STP Workbook; Transparencies
____ Quiz (4.7–4.8): SE page 262

Notes _____

_____

_____

CHAPTER PACING GUIDE	
Day	**Lesson**
1	Assess Ch. 3; 4.1 (all)
2	4.2 (all)
3	4.3 (all)
4	4.4 (all)
5	4.5 (all); 4.6 (begin)
6	4.6 (end); 4.7 (all)
7	**4.8 (all)**
8	Review/Assess Ch. 4

WARM-UP EXERCISES

For use before Lesson 4.8, pages 256–262

State the domain and range of each function.

1. A store sells video cassettes for $5 each. The cost c of buying n video cassettes can be modeled by $c = 5n$.

2. The function $F = 1.8C + 32$ converts Celsius temperatures to Fahrenheit temperatures.

....

DAILY HOMEWORK QUIZ

For use after Lesson 4.7, pages 250–255

1. Solve the equations algebraically.

a. $-7x + 5 = -9$

b. $\dfrac{3}{2}x + 5 = 20$

2. Solve the equations graphically.

a. $2x + 3 = 5$

b. $4x - 2 = 5x$

3. Mike has a lawn-care service. He calculates his monthly cost y of caring for x lawns using the function $y = 1.8x + 25$. In July, his cost was $133. How many lawns did he care for that month?

Lesson 4.8

NAME ______________________ DATE ________

Available as a transparency

Application Lesson Opener

For use with pages 256–262

Recall that a function is a relationship between two quantities, the input and output. In a function, there is exactly one output value for a given input value. A relation is any pairing of input and output values. All functions are relations, but not all relations are functions. Four relations are shown below. For each relation, the input is given in the top row and the output is given in the bottom row. Write *yes* or *no* to tell whether the relation is a function. Explain your answer.

1.

Hours worked	10	12	8	16	14
Amount earned (dollars)	70	84	56	112	98

2.

Age (years)	13	14	13	15	14	15
Height (inches)	50	51	49	56	58	54

3.

Weight of dog (pounds)	31	42	28	48	42	35
Food eaten per day (cups)	3	3.5	2.5	4	4	3

4.

Distance (miles)	90	60	150	120	210
Time (hours)	1.5	1	2.5	2	3.5

NAME _____ DATE _____

Practice A

For use with pages 256–262

Decide whether the graph represents *y* as a function of *x*. Explain your reasoning.

1.
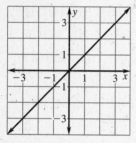

2.
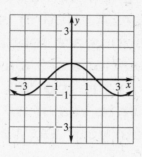

3.
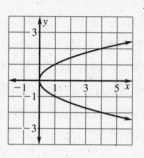

Decide whether the relation is a function. If it is a function, give the domain and the range.

4. Input   Output

    $0 \underset{-6}{\overset{6}{<}}$

    $1 \underset{-5}{\overset{5}{<}}$

5. Input   Output

    1
    2
    3   4
    4

6. Input   Output

    0
    2 → 6
    4 → -3

Evaluate the function when *x* = 3, *x* = 0, and *x* = −2.

7. $f(x) = x$

8. $h(x) = x + 7$

9. $g(x) = x - 2$

10. $g(x) = 3x$

11. $g(x) = 4x - 1$

12. $h(x) = 1.2x$

13. $f(x) = 1.5x - 2$

14. $h(x) = -4x + \frac{1}{2}$

15. $f(x) = \frac{1}{3}x + \frac{2}{3}$

Graph the function.

16. $f(x) = x - 7$

17. $g(x) = 5x$

18. $h(x) = 2x + 1$

19. $g(x) = -4$

20. $f(x) = \frac{1}{2}x - 4$

21. $h(x) = -\frac{2}{5}x + 1$

Decide whether the relation is a function. If it is a function, give the domain and the range.

22.

Input Year	Output Attendance
1996	215
1997	297
1998	412
1999	690
2000	1043

23.

Input Temperature (°F)	Output Date
72°	June 8
74°	June 9
68°	June 10
70°	June 11
70°	June 12

24.

Input Area Code	Output ZIP code
907	99801
916	94203
916	94204
850	32306
217	62706

NAME _____ DATE _____

Practice B

For use with pages 256–262

Decide whether the graph represents *y* as a function of *x*. Explain your reasoning.

1.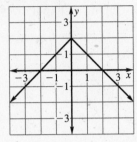

2.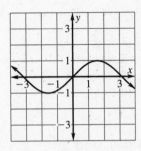

3.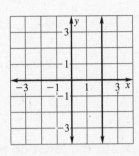

Decide whether the relation is a function. If it is a function, give the domain and the range.

4. Input   Output
$$1 \begin{cases} 7 \\ -7 \end{cases}$$
$$2 \begin{cases} 8 \\ -8 \end{cases}$$

5. Input   Output
$$3 \longrightarrow 2$$
$$5 \longrightarrow 4$$
$$7 \longrightarrow 6$$

6. Input   Output
$$0 \longrightarrow -6$$
$$2 \longrightarrow -4$$
$$4 \longrightarrow -2$$
$$6 \longrightarrow 0$$

Evaluate the function when *x* = 3, *x* = 0, and *x* = −2.

7. $f(x) = 2x - 5$

8. $h(x) = 6x + 2$

9. $g(x) = 2.4x$

10. $f(x) = 0.5x + 12$

11. $h(x) = \frac{2}{3}x - 1$

12. $f(x) = \frac{3}{5}x + 2$

Graph the function.

13. $f(x) = -4x + 3$

14. $g(x) = 2x - 5$

15. $h(x) = -3x - 1$

16. $g(x) = \frac{1}{4}x + 2$

17. $f(x) = -\frac{2}{3}x - 3$

18. $h(x) = -x + 4$

Find the slope of the graph of the linear function *f*.

19. $f(2) = 4, f(0) = 6$

20. $f(1) = 3, f(3) = 7$

21. $f(-2) = 2, f(0) = -4$

22. $f(-1) = -4, f(-2) = 0$

23. $f(-3) = 7, f(2) = -3$

24. $f(-4) = -1, f(3) = 5$

25. *Football Attendance*   The table gives the attendance at a football championship for five consecutive years. Is attendance a function of the number of years since 1993? Why, or why not?

Years since 1993	1	2	3	4	5
Attendance	72,817	74,107	76,347	72,301	68,912

NAME _____ DATE _____

Practice C

For use with pages 256–262

Decide whether the relation is a function. If it is a function, give the domain and the range.

1. Input Output

$$1 \longrightarrow -2$$
$$2 \searrow \nearrow 2$$
$$3 \nearrow \searrow 0$$

2. Input Output

$$1 \longrightarrow 4$$
$$2 \longrightarrow 4$$
$$3 \longrightarrow 5$$
$$4 \longrightarrow 6$$

3. Input Output

$$0 \longrightarrow -6$$
$$2 \longrightarrow -4$$
$$2 \longrightarrow -5$$
$$6 \longrightarrow 0$$

Evaluate the function when $x = 3$, $x = 0$, and $x = -2$.

4. $f(x) = 5 - x$

5. $h(x) = 4x - 1$

6. $g(x) = 12 - 5x$

7. $f(x) = 3.5x + 0.2$

8. $h(x) = \frac{1}{2}x + \frac{3}{2}$

9. $f(x) = 8$

10. $f(x) = 3(x + 2)$

11. $f(x) = \frac{2}{5}(x - 1)$

12. $f(x) = -4(2 - x)$

Graph the function.

13. $f(x) = -3x - 5$

14. $f(x) = -x + 4.5$

15. $f(x) = -\frac{2}{3}x$

16. $g(x) = \frac{3}{4}x + 1$

17. $f(x) = -\frac{3}{2}x + 4$

18. $f(x) = -\frac{1}{6}x + \frac{5}{6}$

Find the slope of the graph of the linear function f.

19. $f(1) = 3, f(0) = 2$

20. $f(2) = -6, f(-5) = 8$

21. $f(-1) = -2, f(-4) = 0$

22. $f(-1) = 4, f(4) = 12$

23. $f(-4) = 7, f(11) = -5$

24. $f(2) = 2, f(8) = -7$

25. Basketball Attendance   The table gives the attendance at a college basketball championship for five consecutive years. Is attendance a function of the number of years since 1993? Why, or why not?

Years since 1993	1	2	3	4	5
Attendance	64,151	23,674	38,540	19,229	47,028

Profits   **In Exercises 26 and 27, use the following information.**

The graph shows the profits of a company from 1994 through 2000.

26. Is the profit a function of the year? Explain.

27. Let $f(t)$ represent the profit in year t. Find $f(1998)$.

28. Challenge   Is $y = |x| + 2$ a function? Why or why not?

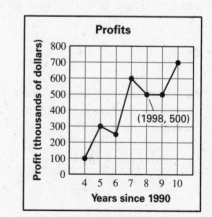

NAME _____ DATE _____

Reteaching with Practice

For use with pages 256–262

GOAL Identify when a relation is a function and use function notation to represent real-life situations.

VOCABULARY

A **relation** is any set of ordered pairs. A relation is a function if for each input there is exactly one output.

Using **function notation,** the equation $y = 3x - 4$ becomes the function $f(x) = 3x - 4$ (the symbol $f(x)$ replaces y). Just as (x, y) is a solution of $y = 3x - 4$, $(x, f(x))$ is a solution of $f(x) = 3x - 4$.

EXAMPLE 1 *Identifying Functions*

Decide whether the relation shown in the input-output diagram is a function. If it is a function, give the domain and the range.

a. Input   Output

$$
\begin{array}{rcl}
1 & \longrightarrow & 4 \\
2 & \longrightarrow & 6 \\
3 & \rightrightarrows & 8 \\
4 & \longrightarrow & 10
\end{array}
$$

b. Input   Output

SOLUTION

a. The relation is not a function, because the input 3 has two outputs: 8 and 10.

b. The relation is a function. For each input there is exactly one output. The domain of the function is the set of input values 1, 2, 3, and 4. The range is the set of output values 5 and 7.

Exercises for Example 1

Decide whether the relation is a function. If it is a function, give the domain and the range.

1. Input   Output

$$
\begin{array}{rcl}
2 & \longrightarrow & 1 \\
4 & \rightleftarrows & 3 \\
  & & 5 \\
8 & \longrightarrow & 7
\end{array}
$$

2. Input   Output

$$
\begin{array}{rcl}
1 & \longrightarrow & 1 \\
2 & \longrightarrow & 4 \\
3 & \longrightarrow & 9 \\
4 & \longrightarrow & 16
\end{array}
$$

3. Input   Output

Copyright © McDougal Littell Inc.
All rights reserved.

Algebra 1
Chapter 4 Resource Book

115

Lesson 4.8

Reteaching with Practice

For use with pages 256–262

EXAMPLE 2 *Evaluating a Function*

Evaluate the function $f(x) = -4x + 5$ when $x = -1$.

SOLUTION

$f(x) = -4x + 5$	Write original function.
$f(-1) = -4(-1) + 5$	Substitute -1 for x.
$= 9$	Simplify.

Exercises for Example 2

Evaluate the function when $x = 3$, $x = 0$, and $x = -2$.

4. $f(x) = 9x + 2$ **5.** $f(x) = 0.5x + 4$ **6.** $f(x) = -7x + 3$

EXAMPLE 3 *Writing and Using a Linear Function*

While on vacation, your family traveled 1800 miles in 5 days.
Your average speed was 360 miles per day.

a. Write a linear function that models the distance that your family traveled each day.

b. Use the model to find the distance traveled after 1.5 days of travel.

SOLUTION

a. **VERBAL MODEL**

$$\boxed{\text{Distance traveled}} = \boxed{\text{Average speed}} \cdot \boxed{\text{Time}}$$

↓

LABELS
Time $= t$ (days)
Average speed $= 360$ (miles per day)
Distance traveled $= f(t)$ (miles)

EQUATION $f(t) = 360t$

b. To find the distance traveled after 1.5 days, substitute 1.5 for t in the function.

$f(t) = 360t$	Write linear function.
$f(1.5) = 360(1.5)$	Substitute 1.5 for t.
$= 540$	Simplify.

Exercises for Example 3

7. Rework Example 3 if your family traveled 2040 miles in 6 days.

8. Rework Example 3 if your family traveled 2660 miles in 7 days.

Lesson 4.8

NAME _____ DATE _____

Quick Catch-Up for Absent Students

For use with pages 256–262

The items checked below were covered in class on (date missed) _____

Lesson 4.8: Functions and Relations

____ **Goal 1:** Identify when a relation is a function. (p. 256)

Material Covered:

____ Student Help: Look Back

____ Example 1: Identifying Functions

Vocabulary:

relation, p. 256

____ **Goal 2:** Use function notation to represent real-life situations. (pp. 257–258)

Material Covered:

____ Student Help: Study Tip

____ Example 2: Evaluating a Function

____ Example 3: Graphing a Linear Function

____ Example 4: Writing and Using a Linear Function

Vocabulary:

function notation, p. 257   graph of a function, p. 257

____ Other (specify) _____

_____

Homework and Additional Learning Support

____ Textbook (specify) _pp. 259–262_____

_____

____ *Reteaching with Practice* worksheet (specify exercises)_____

____ *Personal Student Tutor* for Lesson 4.8

NAME _____ DATE _____

Interdisciplinary Application

For use with pages 256–262

Amelia Earhart's Journey

HISTORY   Amelia Earhart was the first woman to fly across the Atlantic Ocean, both as a passenger and as a pilot, and the first person to fly from Hawaii to California. In 1937, Earhart attempted a flight that no one else at that time had completed. She tried to circumnavigate the globe at the equator. Earhart and her navigator, Fred Newman, disappeared before completing the trip.

On March 17, 1997, the 60th anniversary of Amelia Earhart's journey along the equator in a Lockheed Electra 10E, Linda Finch began World Flight 1997, her own expedition to re-create and complete Amelia Earhart's journey. She completed her journey in 72 days, returning to Oakland on May 28, 1997. Finch covered 26,004 nautical miles on her trip.

1. Write a linear function that models the distance traveled by Linda Finch. (*Hint:* Allow her speed to be measured in nautical miles per day to the nearest hundredth of a nautical mile.)

2. Use your model from Exercise 1 to find the distance traveled after 32 days and 61 days.

3. Graph your model from Exercise 1 and label the points that represent the distances in Exercise 2.

4. The destinations of Amelia Earhart's 1937 flight route are listed in the table below. The table includes the nautical miles to the intended destination for each leg of her journey. Leaving from Oakland, USA, Earhart completed 28 of the 31 legs. Using this information and your model from Exercise 1, find where she could have been on days 32 and 61. (*Note:* The countries' names are listed as they were named in 1937, even though some of them have changed since that time.)

Destination	Nautical miles	Destination	Nautical miles
Burbank, USA	283	Assab, Eritrea	241
Tucson, USA	393	Karachi, India	1627
New Orleans, USA	1070	Calcutta, India	1178
Miami, USA	586	Akyab, Burma	291
San Juan, Puerto Rico	908	Rangoon, Burma	26
Caripito, Venezuela	492	Bangkok, Siam	315
Paramaribo, D. Guiana	610	Singapore	780
Fortaleza, Brazil	1142	Bandoeng, Java	541
Natal, Brazil	235	Surabaya, Java	310
St. Louis, Senegal	1727	Kupeong, Timor	668
Dakar, Senegal	100	Darwin, Australia	445
Gao, Mali	1016	Lae, New Guinea	1012
Fort Lamy, Chad	910	Howland Island, USA	2224
El Fasher, Sudan	610	Honolulu, USA	1648
Khartoum, Sudan	437	Oakland, USA	2090
Massawa, Eritrea	400		

Challenge: Skills and Applications

For use with pages 256–262

For Exercises 1–5, use the relation $x^2 + y^2 = 25$.

1. Find all possible values of y when $x = 3$.

2. Is the relation a function?

3. Make a table of ordered pairs and plot points to graph the relation. Then connect the points using a smooth curve.

4. Explain, using the Vertical Line Test for Functions, how the graph supports your answer to Exercise 2.

5. How can you restrict the range of the relation so that it is a function?

For Exercises 6–7, use the following information.

The height of a ball is a function of the number of seconds since it was thrown into the air.

6. What do you think might be a reasonable domain for this function?

7. What do you think might be a reasonable range for this function?

For Exercises 8–9, use the following information.

The Charity Workers Club has 30 members. One member can complete at most 3 crafts in an hour. The number of crafts completed in one hour meeting is a function of the number of members working on the crafts.

8. What is the domain for this function? What types of numbers make sense in this real-life situation?

9. What is the range for this function? What types of numbers make sense in this real-life situation?

For Exercises 10–12, use the following function.

Two 6-sided number cubes are rolled and the ordered pair generated is the element in the domain. The function rule is to add the two numbers that are rolled. For example, if the first cube shows a 4 and the second shows a 5, the domain element is $(4, 5)$ and the corresponding value in the range is 9.

10. Name five elements of the domain and the five corresponding elements of the range.

11. How many elements are there altogether in the domain of this function?

12. What is the range of this function?

Chapter Review Games and Activities

For use after Chapter 4

The answers to the following exercises will be the *x*-coordinate of an ordered pair listed at the bottom of the page. Place the *y*-coordinate letter on all blanks on the page with the exercise number below them.

$\overline{\quad}$ $\overline{\quad}$ $\overline{\quad}$ $\overline{\quad}$ $\overline{\quad}$ $\overline{\quad}$ $\qquad$ $\overline{\quad}$ $\overline{\quad}$
(12) (5) (3) (3) (13) (11)   (6)   (13)

$\overline{\quad}$ $\overline{\quad}$ $\overline{\quad}$ $\overline{\quad}$ $\overline{\quad}$ $\overline{\quad}$ $\overline{\quad}$ $\overline{\quad}$ $\overline{\quad}$ $\overline{\quad}$ $\overline{\quad}$ $\overline{\quad}$ $\overline{\quad}$
(3) (13) (14) (3) (13) (3) (2) (4)   (12) (7) (10) (9) (4)

1. The *x*-coordinate of $(-5, 3)$.

$\overline{\quad\quad}$
(8)

2. The slope through the points $(-2, 5), (6, -3)$.

$\overline{\quad\quad}$
(3)

3. The *y*-intercept of $3x + 7y = 21$.

$\overline{\quad\quad}$
(12)

4. The slope of $-10x - 5y = 0$.

$\overline{\quad\quad}$ $\overline{\quad\quad}$
(12)   (3)

5. The *y*-coordinate of $(7, -6)$.

$\overline{\quad\quad}$ $\overline{\quad\quad}$
(3)   (8)

6. The *x*-intercept of $-6x + 6y = -6$

$\overline{\quad\quad}$ $\overline{\quad\quad}$
(10)   (9)

7. Evaluate the function $f(x) = 3x - 7$ for $x = 4$.

$\overline{\quad\quad}$ $\overline{\quad\quad}$
(13)   (10)

8. The slope through $(8, 6), (-4, 0)$.

$\overline{\quad\quad}$
(12)

9. The *y*-intercept of $y = \frac{1}{2}x$.

$\overline{\quad\quad}$
(12)

10. The slope of $x - 3y = 15$.

$\overline{\quad\quad}$
(3)

11. The *x*-intercept of $x = 4$.

$\overline{\quad\quad}$
(1)

12. Evaluate the function $g(x) = 5x - 8$ for $x = 2$.

$\overline{\quad\quad}$
(7)

13. The *y*-intercept of $y = 7$.

$\overline{\quad\quad}$
(4)

14. The slope of $y = -\frac{1}{2}x + 4$.

(1, A)	(3, I)	(-2, E)	(0, P)	(2, S)	$(\frac{1}{2}, M)$	(-6, K)
(-5, B)	(7, N)	(-1, T)	$(-\frac{1}{2}, F)$	(5, L)	$(\frac{1}{3}, O)$	(4, G)

NAME _____ DATE _____

Chapter Test A

For use after Chapter 4

Write the ordered pairs that correspond to the points labeled *A, B, C,* and *D* in the coordinate plane.

1.
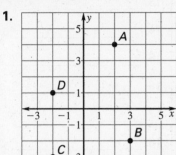

2.
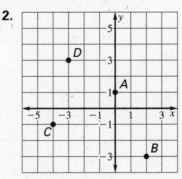

Without plotting the point, tell whether it is in Quadrant I, Quadrant II, Quadrant III, or Quadrant IV.

3. $(7, -10)$

4. $(-4, -8)$

Decide whether the given ordered pair is a solution of the equation.

5. $y - x = 5, (2, 7)$

6. $y + 4 = -2x, (-3, 10)$

Find the *x*-intercept of the graph of the equation.

7. $x + 6y = 7$

8. $4x + y = 3$

Find the *y*-intercept of the graph of the equation.

9. $y - 3x = 4$

10. $2y + x = 8$

Sketch the line that has the given intercepts.

11. *x*-intercept: 1

   y-intercept: 2

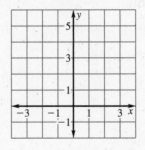

Answers

1. _____
2. _____
3. _____
4. _____
5. _____
6. _____
7. _____
8. _____
9. _____
10. _____
11. Use grid at left.

Algebra 1
Chapter 4 Resource Book

Review and Assess

NAME _____ DATE _____

Chapter Test A

Find the slope of the line passing through the points.

12. $(3, 4), (1, 3)$

13. $(2, 7), (5, 6)$

Find the slope of the line.

14.

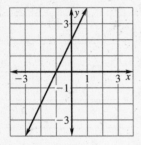

15.

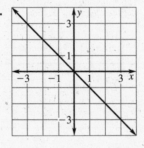

The variables *x* and *y* vary directly. Use the given values to write an equation that relates *x* and *y*.

16. $x = 3, y = 15$

17. $x = 4, y = -16$

Find the slope and *y*-intercept of the graph of the equation.

18. $y = 2x + 5$

19. $y = 5 - 3x$

Solve the equation algebraically.

20. $4x - 3 = 2$

21. $9 = 10 - x$

Decide whether the graphs of the two equations are parallel lines.

22. $y = 2x + 1, \ 2y = 4x + 5$

23. $y = 4x - 3, \ y = -4x + 3$

Decide whether the relation is a function.

24. Input   Output

1	3
2	5
3	7
4	9

25. Input   Output

5	8
10	6
15	4
20	2

12.	_____
13.	_____
14.	_____
15.	_____
16.	_____
17.	_____
18.	_____
19.	_____
20.	_____
21.	_____
22.	_____
23.	_____
24.	_____
25.	_____

Algebra 1
Chapter 4 Resource Book

Review and Assess

Decide whether the given ordered pair is a solution of the equation.

1. $5x + 3y = 2;\ \left(2, -\frac{4}{5}\right)$

2. $\frac{1}{2}x + 4 = 10y;\ \left(2, \frac{1}{2}\right)$

In Questions 3 and 4, use a table of values to graph the equation.

3. $y = \frac{1}{4}x - 2$

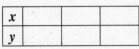

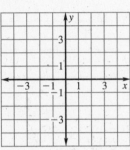

4. $y = 3x - \frac{1}{2}$

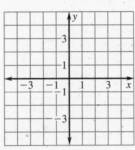

Find the x-intercept of the graph of the equation.

5. $4x + 5y = 8$

6. $3y - 10x = 4$

Find the y-intercept of the graph of the equation.

7. $7 - 12x = 3y$

8. $4y + 9 = 5x$

Find the slope of the line passing through the points.

9. $(-4, 6), (-3, 2)$

10. $(-10, -7), (1, -2)$

Find the value of y so that the line passing through the two points has the given slope.

11. $(6, y), (3, 3), m = \frac{2}{3}$

12. $(8, y), (2, -3), m = \frac{1}{2}$

13. In 1992, a software company had a profit of $30,000,000. In 1998, the company had a profit of $210,000,000. Find the average rate of change of the company's profit in dollars per year.

Answers

1. _____

2. _____

3. Use grid at left.

4. Use grid at left.

5. _____

6. _____

7. _____

8. _____

9. _____

10. _____

11. _____

12. _____

13. _____

Review and Assess

The variables *x* and *y* vary directly. Use the given values to write an equation that relates *x* and *y*.

14. $x = -6, y = 42$

15. $x = \frac{1}{2}, y = 18$

16. You rollerblade at an average speed of 8 miles per hour. The number of miles *m* you rollerblade during *h* hours is modeled by $m = 8h$. Do these two quantities have direct variation?

Write the equation in slope-intercept form. Then graph the equation.

17. $6x - 4y = 3$

18. $2y + 5x = 10$

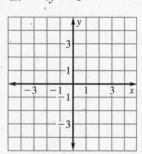

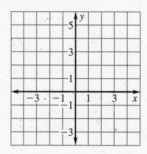

Solve the equation algebraically.

19. $7 - 4x = 5 + 6x$

20. $\frac{3}{5}x + 4 = 10$

Decide whether the graphs of the two equations are parallel lines.

21. $y = 3x + 2, y = \frac{1}{3}x + 4$

22. $3y = 15x + 4, y = 5x + 1$

Evaluate the function when *x* = 3, *x* = 0, and *x* = −2.

23. $f(x) = -\frac{1}{2}x + 3$

24. $h(x) = 5.5x + 4$

25. $g(x) = \frac{1}{8}x - 4$

26. $k(x) = 14 - 4x$

27. Find the slope of the graph of the linear function *f* with $f(0) = 4$ and $f(3) = 13$.

14. _____

15. _____

16. _____

17. _____

18. _____

19. _____

20. _____

21. _____

22. _____

23. _____

24. _____

25. _____

26. _____

27. _____

Review and Assess

Chapter Test C

For use after Chapter 4

Decide whether the given ordered pair is a solution of the equation.

1. $3y + 12x = -4; \left(\frac{1}{5}, -\frac{32}{15}\right)$

2. $8 - 3x + 24y = 0; \left(5, \frac{23}{24}\right)$

In Questions 3 and 4, use a table of values to graph the equation.

3. $3x + 2y = 7$

4. $y = \frac{4}{5}x - \frac{1}{5}$

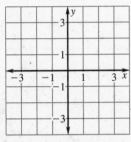

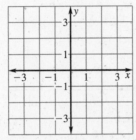

5. Your school biology club is organizing a pancake breakfast to raise $400 for a trip to an aquarium. You decide to charge $2 for each child and $5 for each adult. Write an equation to show the relationship between the number of people and the amount of money raised.

Find the x-intercept of the graph of the equation.

6. $13x + 24y = -5$

7. $-14 + 6y = 7x$

Find the y-intercept of the graph of the equation.

8. $17x + 4y + 10 = 0$

9. $13 + 5y = 15x$

Find the slope of the line passing through the points.

10. $(-5, 5), (-7, -6)$

11. $(7, 12), (4, -13)$

Find the value of y so that the line passing through the two points has the given slope.

12. $(6, y), \left(\frac{3}{2}, \frac{9}{5}\right), m = \frac{2}{3}$

13. $(12, y), \left(-6, \frac{9}{7}\right), m = -\frac{1}{6}$

Answers

1. _____

2. _____

3. Use grid at left. _____

4. Use grid at left. _____

5. _____

6. _____

7. _____

8. _____

9. _____

10. _____

11. _____

12. _____

13. _____

14. In 1990, a restaurant chain had a profit of $45,000. In 1998, the company had a profit of $2,605,000. Find the average rate of change of the chain's profit in dollars per year.

15. Your phone company charges $0.05 per minute for long distance phone calls on the weekend. Write a direct variation model that relates the total cost *x* to the number of minutes *y* spent talking on the phone.

Write the equation in slope-intercept form. Then graph the equation.

16. $18y + 2x - 9 = 0$

17. $3x - 4y = -8$

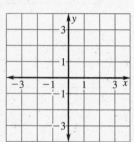
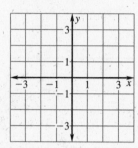

Solve the equation algebraically.

18. $\frac{1}{2}x + \frac{2}{3} = \frac{1}{8}x - \frac{3}{2}$

19. $-4.3 + 5.1x = 5.3x + 7.1$

Decide whether the graphs of the two equations are parallel lines.

20. $2x + 3y = 5,\ 9y + 6x - 1 = 0$

21. $15 + 3x - 10y = 0,\ 30x + 24 = 10y$

Evaluate the function when *x* = 4, *x* = 0, and *x* = −3.

22. $f(x) = 1.5x - 0.4$

23. $h(x) = \frac{3}{8}x + \frac{2}{3}$

24. $g(x) = -14x + 5$

25. $k(x) = 3.2 - 5x$

Find the slope of the graph of the linear function *f*.

26. $f\left(\frac{5}{2}\right) = \frac{7}{2}, f(4) = 6$

14. _____

15. _____

16. _____

17. _____

18. _____

19. _____

20. _____

21. _____

22. _____

23. _____

24. _____

25. _____

26. _____

Review and Assess

1. What is the y-intercept of $3x + 2y = 21$?

(A) $\frac{21}{2}$ **(B)** 14

(C) $\frac{2}{21}$ **(D)** 7

2. What is the equation of the line shown?

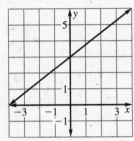

(A) $y = -\frac{3}{4}x + 3$ **(B)** $y = \frac{3}{4}x + 3$

(C) $y = -\frac{4}{3}x + 4$ **(D)** $y = \frac{4}{3}x - 4$

3. Find the slope of the line passing through $(-3, -6)$ and $(7, -2)$.

(A) -2 **(B)** 1

(C) $\frac{4}{5}$ **(D)** $\frac{2}{5}$

4. Find the value of y so that the line passing through $(-3, y)$ and $(4, 4)$ has a slope of -2.

(A) 18   **(B)** 10

(C) 2   **(D)** -6

5. The variables x and y vary directly. When $x = 13$, $y = 52$. Which equation correctly relates x and y?

(A) $y = 13x$ **(B)** $y = \frac{1}{4}x$

(C) $y = 52x$ **(D)** $y = 4x$

6. Find the slope and y-intercept of the graph of $y = \dfrac{5 - x}{10}$.

(A) $m = 10$, y-intercept: 5

(B) $m = -1$, y-intercept: 5

(C) $m = -\frac{1}{10}$, y-intercept: $\frac{1}{2}$

(D) $m = -1$, y-intercept: $\frac{1}{2}$

7. Find the value of $f(x) = 2x - \frac{1}{6}$ when $x = 2$.

(A) $f(2) = \frac{23}{6}$ **(B)** $f(2) = \frac{1}{2}$

(C) $f(2) = \frac{25}{6}$ **(D)** $f(2) = -\frac{23}{6}$

8. The population of a city rises from 100,000 to 226,000 over a ten-year period. Using the points (0, 100,000) and (10, 226,000), find the average rate of change in people per year.

(A) 0.000079 people per year

(B) 12,600 people per year

(C) 0.4375 people per year

(D) 2.29 people per year

Choose the statement that is true about the given numbers.

9.

Column A	Column B
The slope of the line through $(-6, 2)$ and $(4, -2)$	The slope of the line through $(5, 0)$ and $(0, 2)$

(A) The number in column A is greater

(B) The number in column B is greater

(C) The two numbers are equal.

(D) The relationship cannot be determined from the given information.

Review and Assess

Alternative Assessment and Math Journal

For use after Chapter 4

JOURNAL 1. Your classmate Juan missed the lesson on graphing linear equations using slope-intercept form. You attended the class and want to help Juan understand the material he missed. Assume Juan knows how to graph $y = x$. (a) Write an explanation for Juan that describes the y-intercept of a line and what happens to the graph of $y = x$ as you change the y-intercept. Be specific and consider several cases. (b) Write an explanation for Juan that describes the slope m of a line and what happens to the graph of $y = x$ as you change the slope. Be specific and consider several cases.

MULTI-STEP PROBLEM 2. A model for the number of members at a certain health club between 1985 and 2000 is $n = 50t + 300$, where n represents the number of club members and t is the number of years since 1985.

    a. From 1985 to 2000, was the club's membership increasing or decreasing? What was the rate of change (slope)? Describe the meaning of the rate of change in your own words.

    b. Determine the n-intercept for this model and describe its meaning with respect to the club's membership.

    c. In which year were there 1000 club members? Solve algebraically and graphically.

    d. According to the model, how many members did the club have in 1975? Is the result reasonable? Explain.

    e. Use the model to predict how many members the club will have in 2015. Is the result reasonable? Explain.

    f. How can you use the model to approximate the year the club opened? What part of the graph represents that year? Approximate the year the club opened.

3. *Critical Thinking* A model for the number of members at another health club between 1985 and 2000 is $n = -50t + 300$, where n represents the number of club members and t is the number of years since 1985. Compare the second club's membership to the first club's membership over time. Which club is better at attracting more members? Justify your choice.

Alternative Assessment Rubric

For use after Chapter 4

JOURNAL
SOLUTION

1 a, b Complete answers should address these points:

a. • Define y-intercept.

• Explain that a graph of $y = x + k$ is $|k|$ units above or below the graph of $y = x$.

b. • Define slope.

• Explain that a graph of $y = mx$, $m > 0$, slopes up from left to right and gets steeper as m increases.

• Explain that a graph of $y = mx$, $m < 0$, slopes down from left to right and gets steeper as m increases in absolute value.

MULTI-STEP
PROBLEM
SOLUTION

2. a. increasing; 50 members per year; each year there were 50 new club members.

b. 300; in 1985 there were 300 members.

c. 1999

d. -200 members; *Sample answer:* This does not make sense because there can not be a negative number of people. The club probably was not in existence in 1975.

e. 1800; *Sample answer:* The result may or may not be reasonable. The prediction is based on a model that ends in 2000, so it may not be very accurate for 2015.

f. Find the year when the club had no members; the t-intercept; 1979.

3. *Sample answer:* Both clubs had 300 members in 1985, because both graphs have the vertical intercept 300. The club in Exercise 2 is gaining members, because the rate of change is positive. The other club is losing members because the rate of change is negative. Therefore, the club in Exercise 2 is better at attracting members.

MULTI-STEP
PROBLEM
RUBRIC

4 Students complete all parts of the questions accurately. Explanations are logical and clear and demonstrate a grasp of rates of change. Students include a statement that the data is accurate only for the inclusive dates. Graph is completely correct.

3 Students complete the questions and explanations. Solutions may contain minor mathematical errors or misunderstandings. No explanation is given regarding the limits of the data. Graph is sketched accurately.

2 Students complete questions and explanations. Several mathematical errors may occur. Explanations do not fit the model. Graph is incomplete or incorrect.

1 Students' work is very incomplete. Solutions or reasoning are incorrect. Graph is missing or completely inaccurate. Explanations do not follow model.

Review and Assess

Project: Carnival Time

For use with Chapter 4

OBJECTIVE Comparing the expenses and income of a carnival to determine profitability

MATERIALS paper, pencil, graphing calculator or computer (optional)

INVESTIGATION You are on a planning committee for a one-day school carnival. Your committee must decide what activities to include and what to charge. You are considering two options.

Option 1	Option 2
$1.00 admission	no admission
$.25 per ticket	$.50 per ticket

1. List the activities you would like to include. Your list should have:
   - at least 10 activities, including food booths, games, and perhaps rides
   - a good variety that would appeal to a range of people in the community
   - a total cost that stays within the $500 limit you have been given

   You will need to do a rough estimate of the costs. The table shows sample costs for some possible one-day rentals. Also, think about items that people might donate and costs for activities you can make, such as a coin toss.

cotton candy machine	$50
spin art machine	$40
dunk tank	$135
inflatable bungee run	$350

2. Profit equals income (the money taken in) minus expenses. For each option, write an equation to find the profit y of selling x tickets. Assume that 200 people will attend and use $500 as your expenses.

3. Graph the equations. For which number of tickets sold are the two options equal? What is the profit or loss for this number of tickets?

4. Use your graph to decide which option is better. Explain. With the better option, how many tickets must you sell to pay for the activities?

5. Conduct a survey to estimate how many people would attend and the average number of tickets each would buy. How many tickets could you reasonably expect to sell?

6. If your estimate for attendance is not 200, then revise your equations and graph the new equations. Would you still choose the same option? Explain. (*Hint:* Keep in mind the number of tickets you expect to sell.)

7. Use your estimates for attendance and ticket sales and your equations (the revised ones if necessary) to estimate the total income possible from the carnival. Would your carnival make a profit?

PRESENT YOUR RESULTS Write a report to your principal or student council recommending that your school offer or not offer a carnival. Present your activities and discuss why you think they meet the requirements. Show your estimates of income and profit or loss and how you reached them. Provide all supporting evidence including any equations, graphs, and survey data.

Project: Teacher's Notes

For use with Chapter 4

GOALS • Write a linear equation to model a real-life situation.

• Graph and interpret a linear equation in slope-intercept form to solve real-life problems.

• Use a survey to make predictions about a population from a sample.

MANAGING THE PROJECT You may wish to have students work in small groups to simulate a carnival committee. Encourage groups to make collective decisions and to prepare the final report jointly. If necessary, you can break the report into parts and require each student to write the first draft of one part.

You may want to discuss with students how to structure the survey to get useful information. Important points to address are: including specific questions based on the activities being considered, avoiding bias in the wording, and trying to get a sample that is representative of the community.

RUBRIC The following rubric can be used to assess student work.

4 The choice of activities shows attention to variety and to the $500 limit. The student writes and graphs appropriate equations to model the two pricing options, analyzes the models, and modifies them correctly. Students conduct a survey which is unbiased and representative and use it to make reasonable estimates. The report presents an appropriate decision about holding a carnival and makes a clear and convincing case for it that shows all supporting evidence including any expense information gathered, equations, graphs, and survey data.

3 The student chooses at least ten activities and discusses the choice, writes and graphs appropriate equations to model the two options, interprets the graph to answer all the questions, and uses a survey to make estimates. However, the student may not perform all calculations accurately or may not fully address the issues when choosing activities or when designing and predicting from the survey. The report gives and supports an appropriate decision about holding a carnival, but the presentation may not be as convincing as possible.

2 The student chooses and discusses activities, writes and graphs equations to model the two pricing options, answers questions from the graph, and conducts a survey. However, work may be incomplete or reflect misunderstandings. For example, the student may not subtract expenses in the profit equations or may not revise equations based on the survey. The report may indicate a limited grasp of certain ideas or may lack key supporting evidence.

1 Equations, graphs, interpretations of graphs, and predictions using a survey are missing or do not show an understanding of key ideas. The report doesn't give a reasonable decision or fails to support the decision.

Review and Assess

Cumulative Review

For use after Chapters 1–4

Evaluate the expression for the given value of the variable. (1.1)

1. $2x - 1$ when $x = 3$ **2.** $5 - b$ when $b = 2$ **3.** $1.1c$ when $c = -10$

4. $\dfrac{4.2}{x}$ when $x = 2$ **5.** $\dfrac{1}{2}x$ when $x = \dfrac{5}{2}$ **6.** $\dfrac{2}{3} - p$ when $p = \dfrac{1}{9}$

Evaluate the numerical expression. (1.3)

7. $9 \div 3 + 2 \cdot 4$ **8.** $3 \cdot 4^2 \div 12$ **9.** $8[(20 - 4) - 6]$

10. $[15 + (3^2 \cdot 2)] \div 11$ **11.** $\frac{1}{4} \cdot 48 - 3^2$ **12.** $\frac{1}{5}(8 \cdot 10) + 4$

Write the verbal phrase as an algebraic expression. Use *x* for the variable in your expression. (1.5)

13. Six more than a number

14. Difference of fifteen and a number

15. Product of two and a number

Find the sum using the rules of addition. (2.2)

16. $-5 + 7$ **17.** $12 + 30$ **18.** $17 + 0$

19. $-12 + (-8)$ **20.** $4.2 + (-3.1) + 5.4$ **21.** $6.2 + (-1.1) + (-3.4)$

Find the difference. (2.3)

22. $6 - 10$ **23.** $7 - (-6)$ **24.** $-2 - (-4)$

25. $\frac{3}{4} - \frac{5}{4}$ **26.** $-\frac{2}{3} - \frac{1}{3}$ **27.** $\frac{5}{2} - \frac{1}{4}$

Simplify the variable expression. (2.5)

28. $(-6)(-z)$ **29.** $2(-b)(-b)(-b)$ **30.** $-(-y)^2$

31. $|(4)(-z)(-z)(-z)|$ **32.** $-(x^5)(x)$ **33.** $\frac{2}{3}\left(\frac{3}{2}x\right)$

Find the probability of choosing a red marble from a bag of red and white marbles. (2.8)

34. Number of red marbles: 12   **35.** Number of red marbles: 6
      Total number of marbles: 48   Total number of marbles: 22

Solve the equation. (3.1–3.4, 3.6)

36. $15x - 3 = 48$ **37.** $4(32 - 2t) = 144$

38. $-(x - 1) = 2(x - 1)$ **39.** $-2(4.36 - 6.92x) = 13.83x - 2.55$

Cumulative Review

For use after Chapters 1–4

Find the unit rate. (3.8)

40. $1.50 for 4 cans of tuna

41. $2.10 for $2\frac{1}{2}$ gallons of soda

42. $189.00 for working 27 hours

43. 5 liters for 4 servings

Find three different ordered pairs that are solutions of the equation. (4.2)

44. $y = 2x - 1$

45. $y = 9 - 3x$

46. $x = \frac{1}{4}$

47. $y = -1$

Use a table of values to graph the equation. (4.2)

48. $y = -x + 2$

49. $y = 3x + 1$

50. $y = -\frac{1}{2}x + 4$

51. $y = -2$

Find the *x*-intercept and the *y*-intercept of the line. Graph the equation. Label the points where the line crosses the axes. (4.3)

52. $y = x + 3$

53. $2x + 3y = 9$

54. $y = 2x - 1$

55. $x - 4y = 7$

Plot the points and find the slope of the line passing through the points. (4.4)

56. $(2, 3), (4, 1)$

57. $(-1, -2), (-4, -3)$

58. $(3, 5), (-2, 5)$

59. $(7, 2), (-2, -6)$

60. $(1, -1), (1, 7)$

61. $\left(-\frac{1}{2}, \frac{2}{3}\right), \left(\frac{3}{2}, -\frac{4}{3}\right)$

Write the equation in slope-intercept form. Then graph the equation. (4.6)

62. $2y = 4$

63. $2x - 4y = 8$

64. $x + 5y - 1 = 0$

65. $x - y = 0$

Decide whether the graphs of the two equations are parallel lines. Explain your answer. (4.6)

66. $y = 2x + 5, y - 2x = 12$

67. $3x - 4y = 7, y = -\frac{4}{3}x - 5$

In Exercises 68–71, the variables *x* and *y* vary directly. Use the given values to write an equation that relates *x* and *y*. (4.5)

68. $x = 3, y = 6$

69. $x = -2, y = -4$

70. $x = 21, y = 7$

71. $x = -3, y = -3$

Evaluate the function when $x = 3$, $x = 0$, and $x = -1$. (4.8)

72. $f(x) = 15x - 2$

73. $f(x) = 4x$

74. $f(x) = -8x + 1$

75. $f(x) = \frac{1}{2}x - 3$

Review and Assess

ANSWERS

Chapter Support

Parent Guide
Chapter 4

4.1 120 euros converts to \$126; others are correct

4.2 Answers may vary. *Sample answer:* $(0, -2)$, $(5, 0)$, $(-5, -4)$

4.3 $(4, 0)$ and $(0, -7)$ **4.4** $\frac{1}{2}$ **4.5** 9 gallons

4.6 parallel; both have a slope of $\frac{2}{3}$.

4.7 $3x + 6 = 0$; $y = 3x + 6$; -2 **4.8** 10

Prerequisite Skills Review

1. $\frac{1}{4}$, 0.25 **2.** $\frac{3}{5}$, 0.60 **3.** $\frac{7}{100}$, 0.07 **4.** $\frac{57}{50}$, 1.14

5. a. Answers vary.

b.

Input, x	Output, y
0	7
1	5
2	3
3	1
4	-1

c. Domain: 0, 1, 2, 3, 4 Range: 7, 5, 3, 1, -1

6. a. Answers vary.

b.

Input, x	Output, y
-2	-11
-1	-8
0	-5
1	-2
2	1

c. Domain: $-2, -1, 0, 1, 2$
Range: $-11, -8, -5, -1, 1$

7. $\frac{1}{4}$ **8.** -6

Strategies for Reading Mathematics

1. C; D; 4 gal of gasoline and 120 mi **2.** no; yes; yes; yes **3.** no; *Sample answer:* If a point doesn't fall on a grid line, you have to estimate the values.

4. $d = 30g$, where $g \geq 0$; each output, d, is 30 times the corresponding input, g.

Lesson 4.1

Warm-up Exercises

1. $0°$ **2.** $437°F$ **3.** $-40°C$

Quiz

1. 478 mi/h **2.** \$1055/job **3.** 17%
4. about 633 people

Application Lesson Opener
Allow 10 minutes.

1. the number of hours spent babysitting

2. the amount of money earned

3. the point where the axes cross

4. Move right one unit, then move up five units; for babysitting 1 hour, you will make \$5.

5. To get to point B, move right two units, then move up ten units. These numbers represent earning \$10 for 2 hours of babysitting. To get to point C, move right three units, then move up fifteen units. These numbers represent earning \$15 for 3 hours of babysitting. To get to point D, move right four units, then move up twenty units. These numbers represent earning \$20 for 4 hours of babysitting.

6. Yes; *Sample Answer:* You can add the points with coordinates $(5, 25)$ and $(6, 30)$. These points were chosen because they appear to be on a line with A, B, C, and D. These numbers represent earning \$25 for 5 hours of work and \$30 for 6 hours of work.

Graphing Calculator Activities

1. a. The x-value and the y-value are both positive. **b.** The x-value is negative and the y-value is positive. **c.** The x-value and the y-value are both negative. **d.** The x-value is positive and the y-value is negative. **2.** No; check graphs.

Practice A

1. $A(2, 2), B(0, 4), C(-3, -2), D(0, 0)$

2. $A(3, 4), B(-1, 3), C(0, -2), D(4, 0)$

3. $A(-4, 0), B(2, 3), C(1, -4), D(-3, -5)$

4.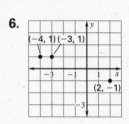

5.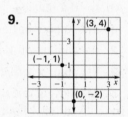

6.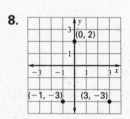

7.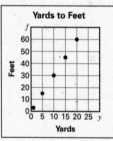

8.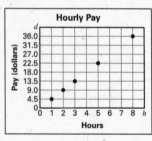

9.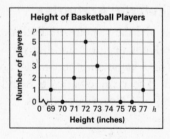

10. Quadrant I   **11.** Quadrant IV

12. Quadrant IV   **13.** Quadrant III

14. Quadrant II   **15.** Quadrant III

16. Quadrant I   **17.** Quadrant II

18.

Hourly Pay

19.

Yards to Feet

20.

Height of Basketball Players

Practice B

1. $A(5, 1)$, $B(0, 4)$, $C(-3, -1)$, $D(0, 0)$

2. $A(4, -5)$, $B(-1, 3)$, $C(0, -2)$, $D(1, 1)$

3. $A(-4, 0)$, $B(2, 3)$, $C(1, -1)$, $D(-3, -5)$

4.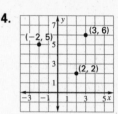

5.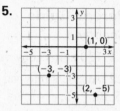

6.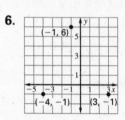

7.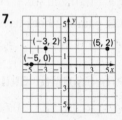

8.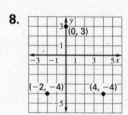

9.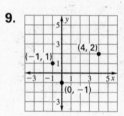

10.

Inches to Centimeters
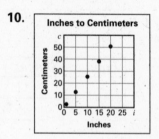

11.

Quiz Grades
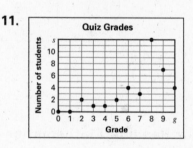

Algebra 1
Chapter 4 Resource Book

Lesson 4.1 *continued*

12.

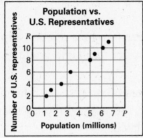

13. The greater the population, the more U.S. representation, about 2 representatives per million in population. (Appears linear)

Practice C

1. $A(5, 1), B(-2, 5), C(-3, -1), D(0, 0)$

2. $A(4, -5), B(-3, 2), C(0, -2), D(1, 1)$

3. $A(-4, 0), B(0, 2), C(1, -1), D(-3, -5)$

4. 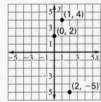 **5.**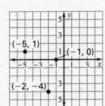

6. 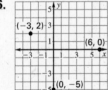 **7.**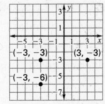

8.  **9.**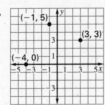

10.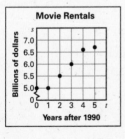

11.

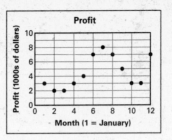

12.

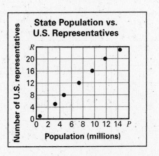

13. The greater the population, the more U.S. representation, about 2 representatives per million in population. (Appears linear)

14. 27; answers may vary

Reteaching with Additional Practice

1. 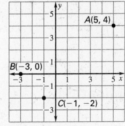 **2.**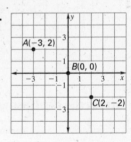

3. 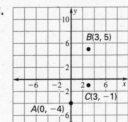 **4.**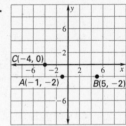

5. 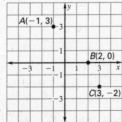 **6.**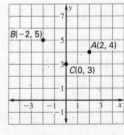

Lesson 4.1 *continued*

7.

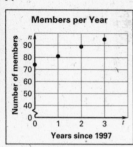

Members per Year

8.

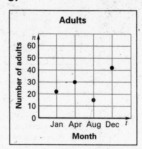

Adults

9.

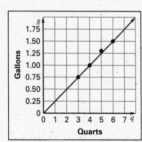

Data is incorrect. Five quarts should be 1.25 gallons.

10.

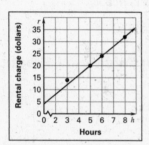

Data is incorrect. Rental charge should be $12 for 3 hours.

Interdisciplinary Application

1.

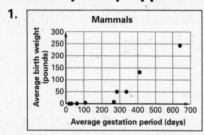

Mammals

2. 243 pounds; 0.0025 pound

3. As the length of the gestation period increases, the birth weight increases.

Challenge: Skills and Applications

1–4. Students can make separate scatter plots for all of the pairs of factors or they can make just two scatter plots (proficiency compared to expenditures per pupil and graduation rate compared to expenditures per pupil.)

1. States with a higher 8th-grade proficiency in 1992 tend to have a higher graduation rate in 1995.

2. There does not seem to be a relationship between expenditures per pupil and graduation rate.

3. The two indicators of success are related to each other. However, success and money spent do not seem to be related.

4. Estimates may vary; about 66%.

Lesson 4.2

Warm-Up Exercises

1. $y = -2x + 10$ **2.** $y = 2x + 1$ **3.** -10
4. 16

Daily Homework Quiz

1. $A(2, 1)$, $C(-2, 2)$, $E(3, -2)$

2.

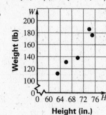

Activity Lesson Opener

Allow 10 minutes.

3. The coordinates of the points on the line make the given equation true.

Practice A

1. b   **2.** b   **3.** b   **4.** a   **5.** b   **6.** b
7. b   **8.** a   **9.** a

10–15. Answers may vary. Sample answers are given below.

10. $(-1, -6)$, $(0, -5)$, $(1, -4)$
11. $(-2, -1)$, $(-2, 0)$, $(-2, 1)$
12. $(-1, 1)$, $(0, 1)$, $(1, 1)$
13. $(-1, 5)$, $(0, 4)$, $(1, 3)$
14. $(-1, -1)$, $(0, -4)$, $(1, -7)$
15. $(-1, 6)$, $(0, 8)$, $(1, 10)$
16. $y = x + 6$ **17.** $y = -x - 2$
18. $y = x - 2$ **19.** $y = 2x - 4$
20. $y = 3x - 1$ **21.** $y = 2x$

Lesson 4.2 *continued*

22. $y = -2x + \frac{1}{2}$ **23.** $y = 3x - 2$

24. $y = -\frac{1}{2}x - \frac{3}{4}$

25. 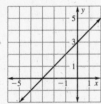   **26.**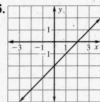

27. 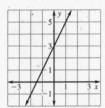   **28.**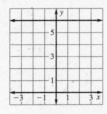

29.    **30.**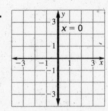

31. 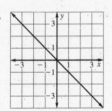   **32.**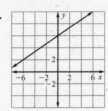

33. 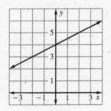   **34.**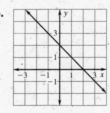

35. 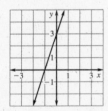   **36.**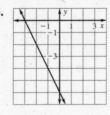

37. $95   **38.** $y = -1.5x + 40$

39.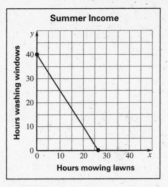

Practice B

1. a **2.** b **3.** b **4.** b **5.** a **6.** a **7.** a

8. b **9.** b **10–15.** Answers may vary.
Sample answers are given below.

10. $(-1, -1), (0, 1), (1, 3)$

11. $(5, -1), (5, 0), (5, 1)$

12. $(-1, -4), (0, -4), (1, -4)$

13. $(-1, 7), (0, 5), (1, 3)$

14. $(-1, 6), (0, 12), (1, 18)$

15. $(-2, -3), (0, -4), (2, -5)$

16. $y = 2x + 6$ **17.** $y = -\frac{1}{4}x - \frac{1}{2}$

18. $y = x + 7$ **19.** $y = \frac{5}{2}x - 2$

20. $y = \frac{3}{5}x - \frac{1}{5}$ **21.** $y = -\frac{1}{2}x$

22. 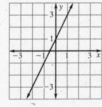   **23.**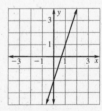

24. 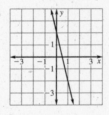   **25.**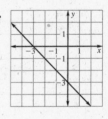

26. 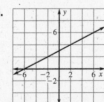   **27.**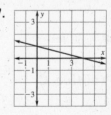

Algebra 1
Chapter 4 Resource Book

A5

Answers

Lesson 4.2 *continued*

Answers

28.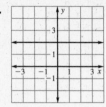

29.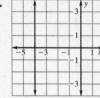

30.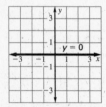

31.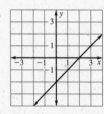

32.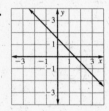

33.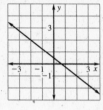

34. $y = -1.5x + 40$

35.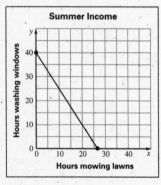

36. 19 hours

37.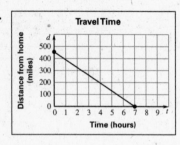

38. 260 miles

10. $(-1, 2), (0, 6), (1, 10)$

11. $(-1, -1), (-1, 0), (-1, 1)$

12. $(-1, 3), (0, 3), (1, 3)$

13. $(-5, 1), (0, 4), (5, 7)$

14. $\left(-2, \frac{5}{2}\right), \left(0, -\frac{1}{2}\right), \left(2, -\frac{7}{2}\right)$

15. $(-1, -1), \left(0, -\frac{1}{4}\right), \left(1, \frac{1}{2}\right)$

16. $y = 7x + 1$ **17.** $y = 2x - 3$

18. $y = -2x + \frac{1}{2}$ **19.** $y = -\frac{3}{4}x - \frac{1}{2}$

20. $y = \frac{6}{5}x - \frac{1}{5}$ **21.** $y = -\frac{2}{3}x$

22.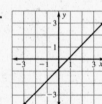

23.

24.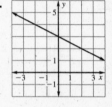

25.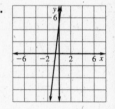

26.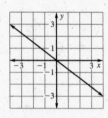

27.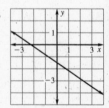

28.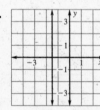

29.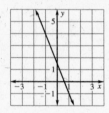

30.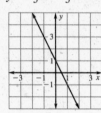

Practice C

1. a   **2.** b   **3.** b   **4.** b   **5.** a   **6.** b   **7.** a

8. b   **9.** b   **10–15.** Answers may vary.
Sample answers are given below.

Lesson 4.2 *continued*

31.

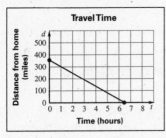

Travel Time

32. 165 miles

33. 6.5 hours **34.** 49 minutes **35.** 58 minutes

36.

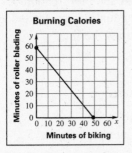

Burning Calories

Reteaching with Additional Practice

1. $(-4, 0)$ is a solution.

2. $(2, 1)$ is not a solution.

3. $(3, 1)$ is a solution.

4. $(-2, 2)$ is not a solution.

5.

Choose x.	−2	−1	0	1	2
Evaluate y.	−10	−7	−4	−1	2

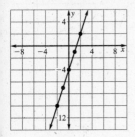

6.

Choose x.	−2	−1	0	1	2
Evaluate y.	0	1	2	3	4

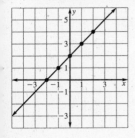

7.

Choose x.	−2	−1	0	1	2
Evaluate y.	9	6	3	0	−3

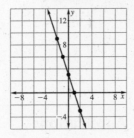

8.

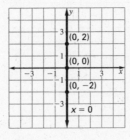

9.

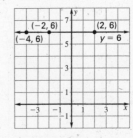

10.

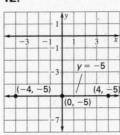

11.

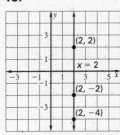

12.

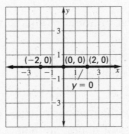

13.

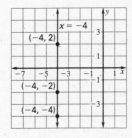

Real-Life Application

1. $y = -x + 45$

2.

x	y
5	40
15	30
25	20
35	10
45	0

3.

Community Service Hours

Answers

Lesson 4.2 *continued*

4. 15 hours   **5.** $y = 45$

6.
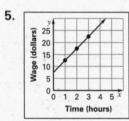
horizontal line

Challenge: Skills and Applications

1. a. $(-4, 3)$ **b.** $(1, -2)$ **c.** $(-5, -1)$
d. $(0, 4)$ **e.** (a, b)

2. charge per hour · number of hours + charge per child · number of children;
charge per hour = \$5;   number of hours = h
charge per child = \$2.50;   number of children = c $5h + 2.50c$

3. *Sample answer:* We know the number of children; $y = 5x + 7.50$; number of hours; amount of money earned

4. *Sample answer:*

x	1	2	3
y	12.50	17.50	22.50

5.
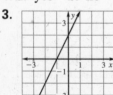
\$27.50

6. 7.50; no; you would not get paid for watching 3 children for 0 hours.

7. *Sample answer:* No; the function from Exercise 3 no longer applies to the situation.

Lesson 4.3

Warm-Up

1. $(3, 0)$ **2.** $(0, -5)$ **3.** 8   **4.** 4

Quiz

1. a. yes   **b.** no   **2.** $y = 2x - 9$
3.
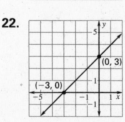
4. a. $(-3, 8)$
b. $(5, 0)$

Activity Lesson Opener

Allow 10 minutes.

3. Students whose coordinates are $(1, 0)$ and $(0, 1)$ should stand.

4. They form the graph of the equation $x + y = 1$.

5. Students whose coordinates are $(0, -2)$ and $(2, 0)$ should stand. They form the graph of the equation $x - y = 2$.

6. *Sample answer:* Find and plot the points where the graph of the equation crosses the axes. Connect the points to draw the line.

Practice A

1. x-intercept: 3; y-intercept: 3
2. x-interecept: 2; y-intercept: -2
3. x-intercept: -1; y-intercept: 3   **4.** 5
5. -6 **6.** 7   **7.** -5 **8.** -15 **9.** 6
10. -4 **11.** 6   **12.** 3   **13.** 3   **14.** -4 **15.** 4

16.
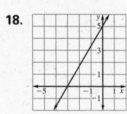
17.
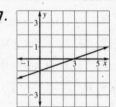

18.
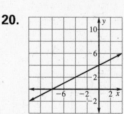
19.
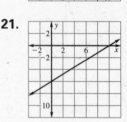

20.
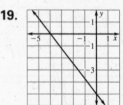
21.
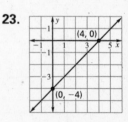

22.
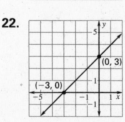
23.

Lesson 4.3 *continued*

24.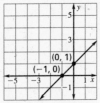

25.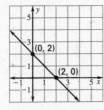

26.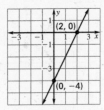

27.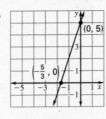

28.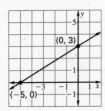

29.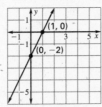

30.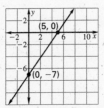

31.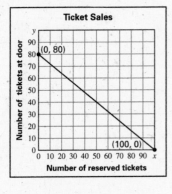

32. *x*-intercept: if no tickets are sold at the door, 100 advanced tickets were sold; *y*-intercept: if no advanced tickets are sold, then 80 tickets were sold at the door.

33.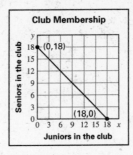

34. *x*-intercept: no seniors, 18 juniors; *y*-intercept: no juniors, 18 seniors

Practice B

1. 5   **2.** 2   **3.** −6   **4.** 6   **5.** −4   **6.** 3

7. −7   **8.** 8   **9.** −$\frac{2}{3}$ **10.** 9   **11.** −8

12. −3

13.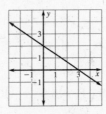

14.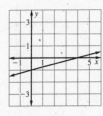

15.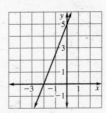

16.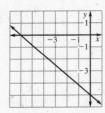

17.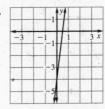

18.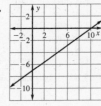

19.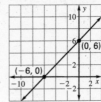

20.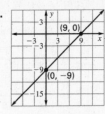

21.
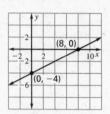

22.
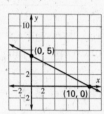

23.
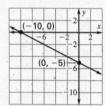

24.
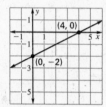

25.

26.
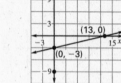

27.
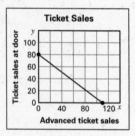

28. $4x + 5.5y = 440$

29.
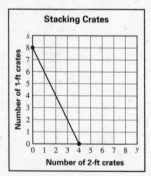

30. *Sample answer:* $(110, 0), (0, 80), (55, 40)$

31.

32. 2 1-foot crates

Practice C

1. *x*-intercept: -9; *y*-intercept: -9
2. *x*-intercept: 2; *y*-intercept: 6
3. *x*-intercept: 4; *y*-intercept: -2
4. *x*-intercept: 6; *y*-intercept: -18
5. *x*-intercept: -6; *y*-intercept: -5
6. *x*-intercept: $-\frac{3}{2}$; *y*-intercept: 2

7.

8.
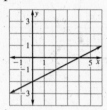

9.
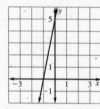

10.
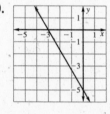

11.
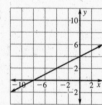

12.
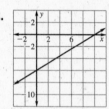

13.
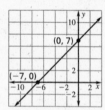

14.
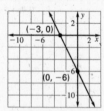

15.
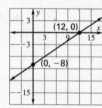

16.
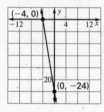

17.
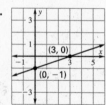

18.
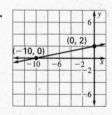

Lesson 4.3 *continued*

19.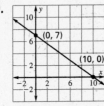

20.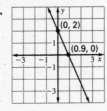

21.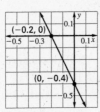

22. $4.5x + 5.5y = 452$

23.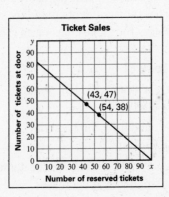

24. 38 tickets

25.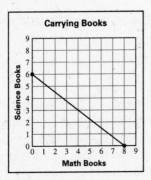

26. No; according to the graph, 5 math books would require you 2.25 science books. You cannot carry part of a book.

27.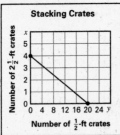

28. $10\frac{1}{2}$-foot crates

1. 6 **2.** 2 **3.** 2 **4.** -6 **5.** -4 **6.** -3

7. The x-intercept is 6 and the y-intercept is 6.

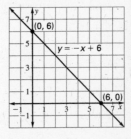

8. The x-intercept is 15 and the y-intercept is -3.

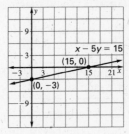

9. The x-intercept is 2 and the y-intercept is 4.

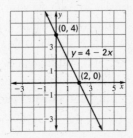

10. The x-intercept is 2 and the y-intercept is -14.

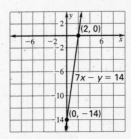

11. The x-intercept is 8 and the y-intercept is 6.

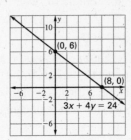

Answers

Lesson 4.3 *continued*

12. The x-intercept is $-\frac{10}{7}$ and the y-intercept is 5.

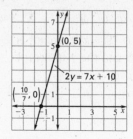

Real Life Application

1.

Amount raised for one shirt	·	Number of shirts sold	+

Amount raised for one sweater	·	Number of sweaters sold	=	Total raised

2. Amount raised for one shirt = 2 (dollars per shirt), Number of shirts sold = x (shirts), Amount raised for one sweater = 4 (dollars per sweater), Number of sweaters sold = y (sweaters), Total raised = 1000 (dollars); $2x + 4y = 1000$

3. 500; $1000 can be raised by selling 500 shirts and no sweaters   **4.** 250; $1000 can be raised by selling 250 sweaters and no shirts

5. no; the x- and y-intercepts have been interchanged.

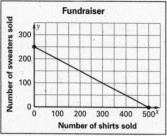

6. *Sample answer:* 100 shirts and 200 sweaters, 150 shirts and 175 sweaters, or 200 shirts and 150 sweaters.

7. *Sample answer:* It should be reasonable to expect to sell 100 shirts and 200 sweaters. While some students will probably not buy a shirt or a sweater, other students may buy several shirts and sweaters for themselves or as gifts.

8. $2x + 4y = 1500$ **9.** x-intercept: 750, $1500 can be raised by selling 750 shirts and no sweaters; y-intercept: 375, $1500 can be raised by selling 375 sweaters and no shirts

10.

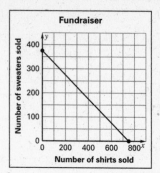

11. *Sample answer:* 100 shirts and 25 sweaters, 150 shirts and 300 sweaters, or 200 shirts and 275 sweaters.   **12.** *Sample answer:* It should be reasonable to expect to sell 100 shirts and 325 sweaters. While some students will probably not buy a shirt or sweater, other students may buy several shirts and sweaters for themselves or as gifts.

Math and History

1. *Sample answer:* about (6.18, 10), about (10, 6.18), (10, 10), (0, 0)

2. *Sample answer:* about (4, 1.02), about (1.02, 4), (8, 8), (0, 0)

3, 4. *Sample answer:* The curves are the same shape, but they get smaller as a decreases.

Challenge: Skills and Applications

1. $5x + 3y = 15$ **2.** $x + 2y = 8$

3. $2x + 3y = 18$ **4.** $5x + 7y = 35$

5. $bx + ay = ab$

6. *Sample Answer:* When $y = 0$, $x = a$. The y-term of the equation drops out, leaving (coefficient of x) $\cdot$ a = right side of equation. Similarly, when $x = 0$, $y = b$. The x-term drops out leaving: (coefficient of y) $\cdot$ b = right side of equation. Therefore, the right side of the equation must be divisible by both a and b. If I use ab for the right side of the equation, then I can make the coefficient of x be b and make the coefficient of y be a and the equations will work.

7. $\frac{3}{4}x + \frac{1}{2}y = 3$ **8.**

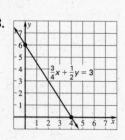

Lesson 4.3 *continued*

9. $(4, 0)$; Steve can make 4 trivets if he does not make any wooden spoons.

10. $(0, 6)$; Steve can make 6 spoons if he does not make any trivets.

11. *Sample answer:* 2 trivets and 3 spoons

Quiz 1

1. 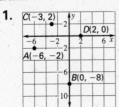   **2.**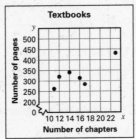

3. Answers may vary. *Sample answer:* $(0, -5)$ and $(1, -2)$ **4.** $y = \frac{1}{2}x - 1$

5. *Sample answer:* $y = 4$ **6.** *x*-intercept, -6; *y*-intercept, -3 **7.**

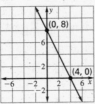

Lesson 4.4

Warm-Up Exercises

1. *x*-coordinate -3; *y*-coordinate 4
2. *x*-coordinate 0; *y*-coordinate -7
3. *x*-coordinate 5; *y*-coordinate 0
4. *x*-coordinate -9; *y*-coordinate -2

Daily Homework Quiz

1. -2 and 4   **2.**

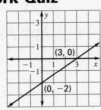

3. a. $c + d = 60$ **b.** $(60, 0)$ and $(0, 60)$; 60 dogs, 0 cats and 0 dogs, 60 cats

Visual Approach Lesson Opener

Allow 5 minutes.

1. 76 ft   **2.** 132 ft   **3.** $\frac{76}{132} = \frac{19}{33}$

4. 44 ft; 191 ft; $\frac{44}{191}$ **5.** 3 ft; 5 ft; $\frac{3}{5}$

Practice A

1. negative   **2.** zero   **3.** positive
4. negative   **5.** undefined   **6.** positive
7. zero   **8.** negative   **9.** positive   **10.** zero
11. negative   **12.** undefined   **13.** 4   **14.** 3
15. -3 **16.** 0   **17.** -2 **18.** undefined
19. 1   **20.** $-\frac{2}{3}$ **21.** $-\frac{6}{7}$ **22.** 0
23. undefined   **24.** $-\frac{5}{2}$ **25.** 3
26. 0   **27.** 8   **28.** 9   **29.** 4   **30.** -7
31. -31 **32.** 1   **33.** -100 **34.** 2.5 miles per minute   **35.** 8 dollars per year
36. 3 gallons per day   **37.** 2 pounds per week
38. 25 miles per gallon   **39.** $-\frac{1}{4}$ inch per month

Practice B

1. negative   **2.** undefined   **3.** positive
4. zero   **5.** negative   **6.** positive   **7.** 6
8. -1 **9.** -3 **10.** 0   **11.** -5
12. undefined   **13.** 2   **14.** $-\frac{3}{2}$ **15.** $\frac{1}{3}$ **16.** 0
17. undefined   **18.** $-\frac{5}{2}$ **19.** 1   **20.** 7   **21.** 14
22. 2   **23.** 4   **24.** -8 **25.** 11 feet per second
26. 13 dollars per week   **27.** 0.8 cent per year
28. -6 dollars per year
29. -139.5 home runs per year
30. 134 home runs per year   **31.** 828 home runs per year   **32.** The six-year rate of change $\left(274\frac{1}{6}\right)$ is the average of the three two-year rates of change.

Practice C

1. $\frac{1}{2}$ **2.** $\frac{9}{10}$ **3.** -1 **4.** -2 **5.** 1
6. undefined   **7.** -3 **8.** $\frac{1}{3}$ **9.** 1   **10.** -12
11. undefined   **12.** -2 **13.** -8 **14.** 3
15. 40   **16.** -2 **17.** $\frac{3}{2}$ **18.** 1
19. 90 kilometers per hour   **20.** 160 dollars per month   **21.** 10 meters per second   **22.** 3 inches per minute

Lesson 4.4 *continued*

23. 750 books per year **24.** −25 births per year
25. 0.24 million students per year
26. 0.11 million students per year
27. −2 million voters per year
28. −0.58 million unemployed laborers per year

29. $\dfrac{7}{x}$; 7 **30.** $-\dfrac{6}{x}$; 2

Reteaching with Practice

1. $m = \dfrac{3}{7}$

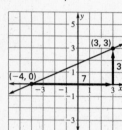

2. $m = \dfrac{-4}{3} = -\dfrac{4}{3}$

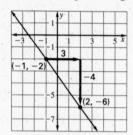

3. $m = 1$

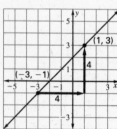

4. The slope is undefined.

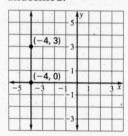

5. The slope is undefined.

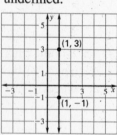

6. The slope is 0.

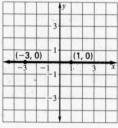

7. The slope is 0.

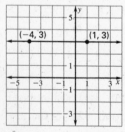

8. The slope is undefined.

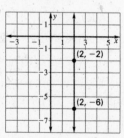

9. The slope is 0.

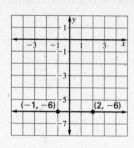

10. The rate of change is $\dfrac{1{,}731{,}000}{3} = 577{,}000$ people per year.

11. The rate of change is $\dfrac{-0.5}{6} \approx -0.08$ million registered motorcycles per year.

Cooperative Learning Activity

1. Answers may vary. **2.** *Sample answer:* Increasing the run of the ramp or decreasing the rise of the ramp will decrease the overall slope.

3. *Sample answer:* To prevent the construction of a ramp that is too steep for the disabled to use.

Interdisciplinary Application

1. 0.0875 dollar per year
2. 0.1928 dollar per year **3.** from 1975 to 1980
4. *Sample answer:* $5.50; comparisons may vary.
5.

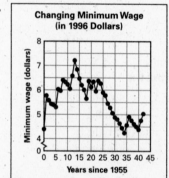

Changing Minimum Wage
(in 1996 Dollars)

Challenge: Skills and Applications

1. $\dfrac{b - 5}{a - 2}$ or $\dfrac{5 - b}{2 - a}$ **2.** undefined **3.** 0
4. −1 **5.** *Sample answer:* $a = 1$ and $b = 4$
6. *Sample answer:* $a = 3$ and $b = 4$ **7.** $a < 2$
8. $a > 2$ **9.** 4 **10.** 2 **11.** −8 **12.** 3

Lesson 4.5 *continued*

Lesson 4.5

Warm-Up
1. 2; 0   **2.** $-3, 0$ **3.** $d = 60t$

Quiz
1. a. 1   **b.** $\frac{1}{3}$

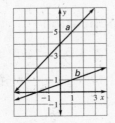

2. 2   **3.** $6/hr

Activity Lesson Opener

Allow 10 minutes.

1–5. Answers will depend upon type of ruler used.

6. The constant of variation is the thickness of the spine of one algebra book.

Practice A

1. yes; $k = 1, m = 1$ **4.** $k = 2, m = 2$

2. no

3. yes; $-3; -3$

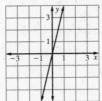

5. $k = 3, m = 3$ **6.** $k = 4, m = 4$

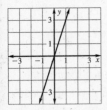

7. $k = -1, m = -1$ **8.** $k = -2, m = -2$

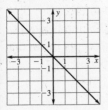

9. $k = \frac{1}{3}, m = \frac{1}{3}$ **10.** $y = 2x$ **11.** $y = 3x$

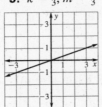

12. $y = 2x$ **13.** $y = 4x$ **14.** $y = -x$

15. $y = -5x$ **16.** $y = 7x$ **17.** $y = \frac{1}{2}x$

18. $y = \frac{1}{4}x$ **19.** $C = 2\pi r$ **20.** $G = 6M$

21.

Total pay, p	$20	$15	$30	$25
Hours worked, h	4	3	6	5
Ratio	5	5	5	5

22. $p = 5h$ **23.** $40

Practice B

1. yes; $k = 5, m = 5$ **2.** no

3. yes; $k = \frac{1}{2}, m = \frac{1}{2}$

4. $k = 3, m = 3$ **5.** $k = -4, m = -4$

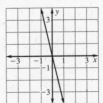

6. $k = 0.5, m = 0.5$ **7.** $k = -0.2, m = -0.2$

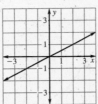 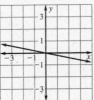

8. $k = -\frac{2}{3}, m = -\frac{2}{3}$ **9.** $k = \frac{1}{4}, m = \frac{1}{4}$

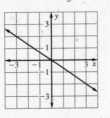 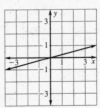

10. $y = 8x$ **11.** $y = \frac{1}{4}x$ **12.** $y = 3x$

13. $y = -4x$ **14.** $y = -\frac{1}{4}x$ **15.** $y = 1.6x$

Lesson 4.5 *continued*

16. $y = -1.5x$ **17.** $y = 0.3x$ **18.** $y = 1.6x$

19. $y = 64$ **20.** $y = 5$ **21.** $y = 32$

22. $y = 15$ **23.** $C = 2\pi r$; $C = 5\pi$ when
$r = 2.5$ **24.** $S = 8.2h$; $S = \$328$ when $h = 40$

25. $G = 6M$; $G = 51$ gallons when
$M = 8.5$ minutes

26. $F = s$; $F = 4$ pounds when $s = 4$ inches

Practice C

1. yes; $k = -\frac{1}{3}, m = -\frac{1}{3}$ **2.** no

3. yes; $k = 5, m = 5$

4. $k = 6, m = 6$ **5.** $k = -10, m = -10$

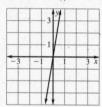 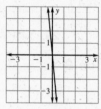

6. $k = -0.6, m = -0.6$ **7.** $k = 1.2, m = 1.2$

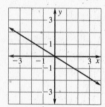 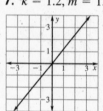

8. $k = \frac{1}{6}, m = \frac{1}{6}$ **9.** $k = -\frac{4}{3}, m = -\frac{4}{3}$

10. $y = 9x$ **11.** $y = -11x$ **12.** $y = 7x$

13. $y = -\frac{2}{3}x$ **14.** $y = -\frac{2}{5}x$ **15.** $y = \frac{6}{5}x$

16. $y = -\frac{3}{2}x$ **17.** $y = 0.9x$ **18.** $y = -2x$

19. $x = 4$ **20.** $x = -3$ **21.** $x = 16$

22. $x = -32$

23. $E = 2I$; $E = 6$ volts when $I = 3$ amperes

24. $C = 2\pi r$; $r = 5$ inches when $C = 10\pi$

25. $G = 6M$; $M = 10$ minutes when $G = 60$
gallons **26.** $F = \frac{5}{4}s$; $s = 16$ inches when $F = 20$
pounds

Reteaching with Additional Practice

1. $y = 5x$ **2.** $y = 0.5x$ **3.** $y = x$

4. $y = -0.2x$ **5.** $y = 2x$ **6.** $y = \dfrac{-x}{3}$

7. $M = 45.6$ **8.** $E \approx 163.2$

9. a. $E = \frac{160}{139}V$ **b.** $V \approx 169.4$

Real-Life Application

1. $c = 1.34g$ **2.** $\$25.46$ **3.** The constant of
variation. **4.** $1.19 = \dfrac{c}{g}$ or $1.19g = c$

5. $\$21.42$

Challenge: Skills and Applications

1. direct variation; $k = 8, A = 8L$ **2.** no direct
variation **3.** direct variation; $k = 400$; $V = 400d$

4. *Sample answer:* Yes, there is a relationship.
The volume is a product of a constant depth time
the square of the side. **5.** about 514 calories

6. $\$30.00$ **7.** yes; 10

Lesson 4.6

Warm-Up

1. $-6; 0$ **2.** $3; -7$ **3.** $4; -5$ **4.** $\frac{7}{3}; -\frac{4}{3}$

Quiz

1. a. $\frac{1}{4}; \frac{1}{4}$ **b.** $-2.5; -2.5$ **2.** yes

3. a. $y = \frac{18}{5}x$ **b.** $y = -1.7x$

4. 0.4 min, or 24 sec

Application Lesson Opener

Allow 15 minutes.

1. slope $= 3$, y-intercept $= 6$, no x-intercept
shown; *Sample answer:* The slope gives the cost
for one pool pass, $\$3$. The y-intercept could be the
fee to sign up to buy pool passes, $\$6$.

2. slope $= \frac{1}{2}$, y-intercept $= 0$, x-intercept $= 0$;
Sample answer: The slope gives the number of
pizzas needed per person. The intercepts mean that
0 pizzas should be ordered for 0 people.

Lesson 4.6 *continued*

3. slope = 1.25, *y*-intercept = 0, *x*-intercept = 0; *Sample answer:* The slope gives the cost per serving, $1.25. The intercepts mean that 0 servings cost $0.

4. slope = −2.5, *y*-intercept = 60, *x*-intercept = 24; *Sample answer:* The slope gives the change in the number of minutes remaining per day, −2.5. The *y*-intercept means that the monthly pre-paid plan includes 60 minutes. The *x*-intercept means that the pre-paid minutes were used up by the 24th day of the month.

5. *Sample answer:* First plot the point for the *y*-intercept, then use the slope to find the second point. Connect the points to draw the line.

Graphing Calculator Activities

1. a. 3   **b.** −5   **c.** $\frac{2}{3}$

2. a. *Sample answer:* $3x + 1, 3x - 1$

b. *Sample answer:* $-5x + 2, -5x - 1$

c. *Sample answer:* $\frac{2}{3}x + 1, \frac{2}{3}x - 2$

Practice A

1. 2; 2   **2.** $-\frac{1}{3}$; −3   **3.** $\frac{1}{2}$; −2   **4.** 7; 3

5. 5; −1   **6.** 0; 7   **7.** −2; 0   **8.** $\frac{1}{2}$; $\frac{5}{2}$ **9.** 2; $\frac{3}{2}$

10.

11.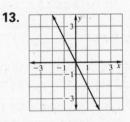

12.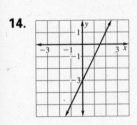

13.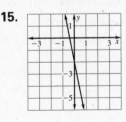

14.

15.

16.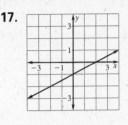

17.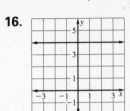

18.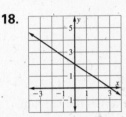

19.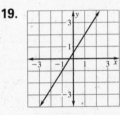

20.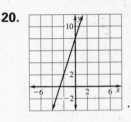

21.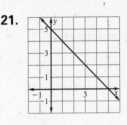

22. yes   **23.** no   **24.** no   **25.** yes

26.

Week, *t*	0	1	2	3	4	5	6	7
Laps, *l*	4	5	6	7	8	9	10	11

27.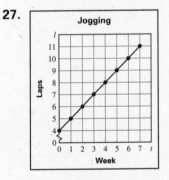

28. $m = 1$; the slope represents the rate at which Howard increases his laps each week.

29.

Minutes, *t*	Cost, *c*
1	$.50
2	$.60
3	$.70
4	$.80
5	$.90
6	$1.00

30.

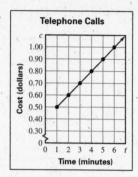

Telephone Calls

31. $m = 0.10$; the slope represents the amount the cost increases with each minute.

Practice B

1. 7; 1 **2.** $-3; -4$ **3.** 0; -4 **4.** 2; 3.2

5. $\frac{1}{4}; \frac{3}{4}$ **6.** 3; 8

7.

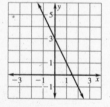

8.

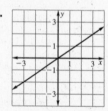

9.

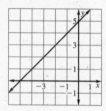

10.

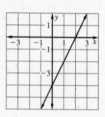

11.

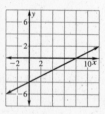

12.

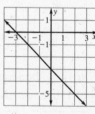

13.

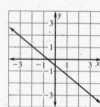

14.

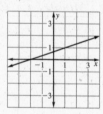

15.

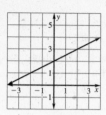

16.

17.

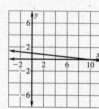

18.

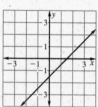

19. no **20.** yes **21.** no **22.** yes **23.** no **24.** yes

25.

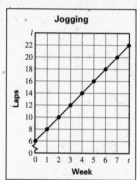

Jogging

26. $m = 2$; the slope represents the rate at which Howard increases his laps each week.

27.

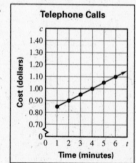

Telephone Calls

28. $m = 0.05$; the slope represents the amount the cost increases with each minute.

29. $m = -1$; the slope represents how much weight the wrestler loses each week.

30. w-intercept = 190; the w-intercept represents the wrestler's starting weight.

Lesson 4.6 *continued*

Practice C

1. $-5; -8$ **2.** $-\frac{3}{4}; 0$ **3.** $-6; 1.8$ **4.** $\frac{1}{2}; 0.8$

5. $\frac{3}{8}; -\frac{7}{8}$ **6.** $\frac{5}{9}; -\frac{5}{3}$

7. 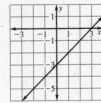 **8.**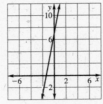

9. 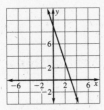 **10.**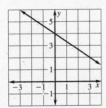

11. 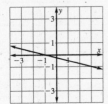 **12.**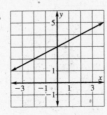

13. 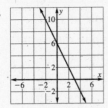 **14.**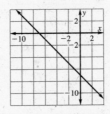

15. 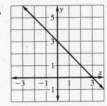 **16.**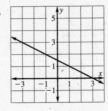

17. 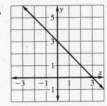 **18.**

19. yes **20.** no **21.** yes **22.** no **23.** yes **24.** no

25. 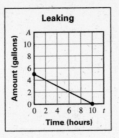 **26.**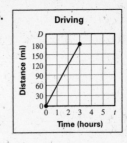

27. 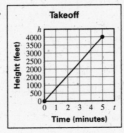 **28.**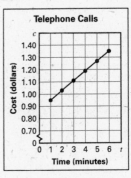

29. $m = 0.08$; the slope represents the amount the cost increases with each minute.

30. -1; the wrestler lost an average of 1 pound per week. **31.** 190; the wrestler originally weighed 190 pounds.

32. The graph would be steeper.

Reteaching with Practice

Slope-intercept form	Slope	y-intercept
1. $y = -3x + 0$	$m = -3$	$b = 0$
2. $y = -x + 5$	$m = -1$	$b = 5$
3. $y = -3x + 5$	$m = -3$	$b = 5$
4. $y = -\frac{1}{3}x + \frac{7}{3}$	$m = -\frac{1}{3}$	$b = \frac{7}{3}$
5. $y = 0x + 2$	$m = 0$	$b = 2$
6. $y = -\frac{1}{4}x + 1$	$m = -\frac{1}{4}$	$b = 1$

7. $y = -3x$ and $3x + y = 5$; they have the same slope, -3.

8. $y = 6x$ **9.** $y = -\frac{1}{3}x + 1$

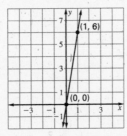 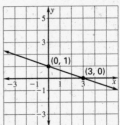

Answers

Lesson 4.6 *continued*

Answers

10. $y = -5x + 4$

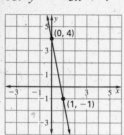

11. $y = -\frac{1}{3}x + 2$

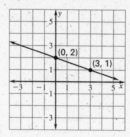

12. $y = -2x + 9$

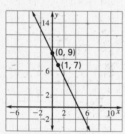

13. $y = -\frac{1}{2}x - 4$

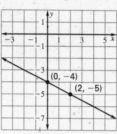

14. a. $w = 1.75h + 4$

b. The slope is 1.75 and the y-intercept is 4.

c. The slope represents the hourly rate.

d.

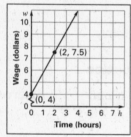

15. a. $w = 1.25h + 6$

b. The slope is 1.25 and the y-intercept is 6.

c. The slope represents the hourly rate.

d.

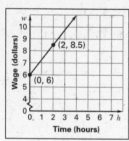

Interdisciplinary Application

1. $\frac{2207}{6600}$ **2.** $y = \frac{2207}{6600}x + 18{,}000$; the slope means that for every 2207 feet gained in altitude, 6600 feet is traveled horizontally; the y-intercept is the altitude of the base camp in feet.

3.

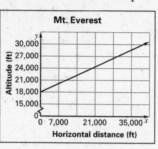

4. 25,062 ft

Challenge: Skills and Applications

1. $y = -\dfrac{A}{B}x + \dfrac{C}{B}$ **2.** $-\dfrac{A}{B}, \dfrac{C}{B}$

3. $m = \frac{5}{3}, b = -5$ **4.** $m = \frac{2}{7}, b = 2$

5. $m = -\frac{3}{2}, b = -3$ **6.** $m = -\frac{4}{5}, b = \frac{7}{5}$

7. $(4, 0)$ and $(0, -3)$ **8.** $(0, -3)$ and $(4, 0)$

9.

Equation	Points plotted
	With intercepts method
$2x + 5y = 10$	$(5, 0)$ and $(0, 2)$
$4x + 5y = 10$	$\left(\frac{5}{2}, 0\right)$ and $(0, 2)$
$7x + 2y = -14$	$(-2, 0)$ and $(0, -7)$
$4x + 6y = -12$	$(-3, 0)$ and $(0, -2)$
$-9x + 4y = -36$	$(4, 0)$ and $(0, -9)$

Equation	Points plotted
	With slope-intercept method
$2x + 5y = 10$	$(0, 2)$ and $(5, 0)$
$4x + 5y = 10$	$(0, 2)$ and $(5, -2)$
$7x + 2y = -14$	$(0, -7)$ and $(2, -14)$
$4x + 6y = -12$	$(0, -2)$ and $(3, -4)$
$-9x + 4y = -36$	$(0, -9)$ and $(4, 0)$

10. $C = AB$

Lesson 4.6 *continued*

Quiz 2

1. $m = -\frac{1}{2}$ **2.** $y = 7$ **3.** undefined

4. $y = \frac{3}{4}x$

5. constant of variation, **6.** $y = x + 10$
$-\frac{1}{2}$; slope $-\frac{1}{2}$.

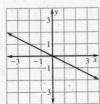 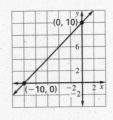

7. The lines are parallel because they have the same slope; $m = \frac{1}{2}$.

Lesson 4.7

Warm-Up Exercises
1. -4 **2.** -1 **3.** 2 **4.** -3 **5.** -5

Daily Homework Quiz

1. $2; -4$
2. $y = -x - 3$

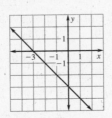

3. Yes; both have slope 2.
4. a square; 36 square units

Graphing Calculator Lesson Opener
Allow 15 minutes.

1. $3x + 36 = 0$ **2.** $y = 3x + 36$
3. Check graphs. **4.** -12; it satisfies the equation. **5.** -12; the algebraic solution is the same as the x-intercept. The x-intercept is the solution of the equation.

6a. 1. $-\frac{1}{2}x - 5 = 0$; **2.** $y = -\frac{1}{2}x - 5$;
3. Check graphs. **4.** -10; it satisfies the equation. **5.** -10; the algebraic solution is the same as the x-intercept. The x-intercept is the solution of the equation.

6b. 1. $\frac{1}{4}x - \frac{5}{4} = 0$; **2.** $y = \frac{1}{4}x - \frac{5}{4}$; **3.** Check graphs. **4.** 5; it satisfies the equation. **5.** 5; The algebraic solution is the same as the x-intercept. The x-intercept is the solution of the equation.

7. *Sample answer:* The x-intercept of a linear equation written in $ax + b = 0$ form is the solution to the equation.

Practice A

1. $(4, 0)$ **2.** $(5, 0)$ **3.** $(3, 0)$ **4.** B **5.** D
6. A **7.** E **8.** F **9.** C
10. $5x - 5 = 0; y = 5x - 5$
11. $-3x - 6 = 0; y = -3x - 6$
12. $4x - 8 = 0; y = 4x - 8$
13. $4x - 9 = 0; y = 4x - 9$
14. $11x - 5 = 0; y = 11x - 5$
15. $8x + 3 = 0; y = 8x + 3$
16. -1 **17.** 3 **18.** 4 **19.** 6 **20.** 15
21. -40 **22.** $-\frac{4}{5}$ **23.** 3 **24.** 2 **25.** -3
26. $-\frac{1}{3}$ **27.** 1 **28.** $\frac{9}{5}$ **29.** $-\frac{4}{3}$ **30.** $-\frac{3}{4}$
31. 1995 **32.** 1999

Practice B

1. F **2.** C **3.** B **4.** D **5.** E **6.** A
7. $5x - 4 = 0, y = 5x - 4$
8. $-3x - 7 = 0, y = -3x - 7$
9. $4x - 20 = 0, y = 4x - 20$
10. $6x - 9 = 0, y = 6x - 9$
11. $-16x + 5 = 0, y = -16x + 5$
12. $-x + 12 = 0, y = -x + 12$ **13.** 2
14. 4 **15.** -3 **16.** $-\frac{1}{5}$ **17.** -2
18. -4 **19.** -10 **20.** 1 **21.** 7 **22.** -1
23. 2 **24.** $\frac{1}{4}$ **25.** 6 **26.** 3 **27.** -12
28. 160 **29.** 840 **30.** 10 cm **31.** 2001

Practice C

1. $5x - 6 = 0, y = 5x - 6$
2. $-3x + 16 = 0, y = -3x + 16$
3. $6x - 8 = 0, y = 6x - 8$
4. $3.4x - 9 = 0, y = 3.4x - 9$
5. $-11.5x + 5 = 0, y = -11.5x + 5$
6. $-x - 3 = 0, y = -x - 3$

Answers

Lesson 4.7 *continued*

7. -2 **8.** $-\frac{9}{2}$ **9.** 1 **10.** 7 **11.** -21

12. -4 **13.** 10 **14.** 1 **15.** $\frac{8}{3}$ **16.** -1

17. $-\frac{5}{2}$ **18.** 16 **19.** 1 **20.** 1 **21.** -9

22. $-\frac{11}{4}$ **23.** -2 **24.** -2 **25.** $\frac{1}{2}$ **26.** $\frac{86}{3}$

27. $\frac{42}{5}$ **28.** 950 **29.** 3 in. **30.** 2008

31. 1993 **32.** 2010

Reteaching with Additional Practice

1. 1

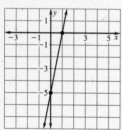

2. -5

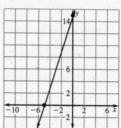

3. -3

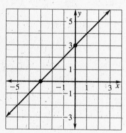

4. -2

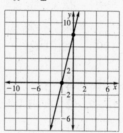

5. 4

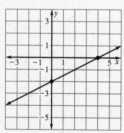

6. 1

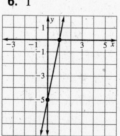

7. The United States will have a consumer price index of 180.4 in the year 2002.

Real-Life Application

1. 2002 **2.** *Sample answer:* Algebraically is more convenient than graphically.

3.

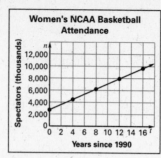

Estimate for 2005: 9200 thousand

4. 2048 **5.** 2001

6.

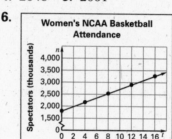

Estimate for 2004: 3100 thousand

Challenge: Skills and Applications

1. 2.325 **2.** $y = 2.325x + 7.6$

3. 12.25 million **4.** The model was off by about a million music videos.

5. $2.325x - 17.4 = 0$ **6.** Estimates may vary. About 7.5; 25 million music videos should be shipped in 1999. **7.** $2.325x - 32.4 = 0$

8. Estimates may vary. About 13.9; 40 million music videos should be shipped in 2005.

9. 92.85 **10.** $y = 92.85x + 407.5$

11. $92.85x - 592.5 = 0$ **12.** Estimates may vary. About 6.4; a billion CD albums should be shipped in 1998.

Lesson 4.8

Warm-Up
1. Domain: whole numbers; Range: nonnegative multiples of 5
2. Domain: real numbers; Range: real numbers

Quiz
1. a. 2 **b.** 10

2.

a. 1 **b.** -2

Lesson 4.8 *continued*

3. 60 lawns

Activity Lesson Opener

Allow 10 minutes.

1. Yes; for each input there is exactly one output. **2.** No; there are two output values for the input values 13, 14, and 15. **3.** No; there are two output values for the input value 42. **4.** Yes; for each input there is exactly one output.

Practice A

1. yes; any vertical line will pass through the graph only once. **2.** yes; any vertical line will pass through the graph only once **3.** no; there are vertical lines that pass through two points on the graph **4.** no **5.** yes; the domain is 1, 2, 3, 4 and the range is 4. **6.** yes; the domain is 0, 2, 4 and the range is $-3, 6$. **7.** 3; 0; -2

8. 10; 7; 5 **9.** 1; -2; -4 **10.** 9; 0; -6

11. 11; -1; -9 **12.** 3.6; 0; -2.4

13. 2.5; -2; -5 **14.** $-\frac{23}{2}; \frac{1}{2}; \frac{17}{2}$ **15.** $\frac{5}{3}; \frac{2}{3}; 0$

16. 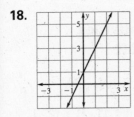 **17.**

18. 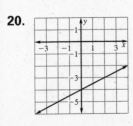 **19.**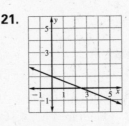

20. 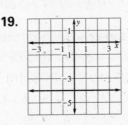 **21.**

22. yes; the domain is 1996, 1997, 1998, 1999, 2000 and the range is 215, 297, 412, 690, 1043.
23. no **24.** no

Practice B

1. yes **2.** yes; any vertical line will pass through the graph only once. **3.** no; the input value of 2 has many output values **4.** no **5.** yes; the domain is 3, 5, 7 and the range is 2, 4. **6.** yes; the domain is 0, 2, 4, 6 and the range is $-6, -4, -2, 0$.

7. 1; -5; -9 **8.** 20; 2; -10 **9.** 7.2; 0; -4.8

10. 13.5; 12; 11 **11.** 1; -1; $-\frac{7}{3}$ **12.** 3.8; 2; $\frac{4}{5}$

13. 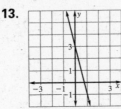 **14.**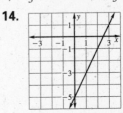

15. 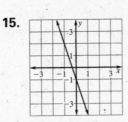 **16.**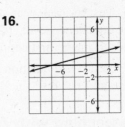

17. 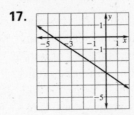 **18.**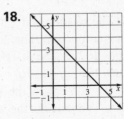

19. -1 **20.** 2 **21.** -3 **22.** -4 **23.** -2

24. $\frac{6}{7}$ **25.** Yes; for each year since 1993 there is exactly one attendance level.

Practice C

1. yes; the domain is 1, 2, 3 and the range is $-2, 0, 2$ **2.** yes; the domain is 1, 2, 3, 4 and the range is 4, 5, 6. **3.** no **4.** 2; 5; 7

5. 11; -1; -9 **6.** -3; 12; 22

7. 10.7; 0.2; -6.8 **8.** 3; $\frac{3}{2}; \frac{1}{2}$ **9.** 8; 8; 8

10. 15, 6, 0 **11.** $\frac{4}{5}, -\frac{2}{5}, -\frac{6}{5}$ **12.** 4, -8, -16

13. 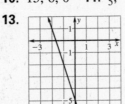 **14.**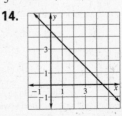

15. 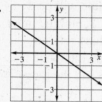   **16.**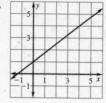

17. 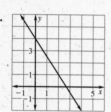   **18.**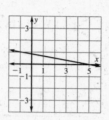

19. 1   **20.** -2 **21.** $-\frac{2}{3}$ **22.** $\frac{8}{5}$ **23.** $-\frac{4}{5}$ **24.** $-\frac{3}{2}$

25. yes; for each year since 1993 there is exactly one attendance level.   **26.** yes; no 2 ordered pairs have the same first coordinate.   **27.** \$500,000

28. yes; it is a function because it passes the Vertical Line Test.

Reteaching with Additional Practice

1. The relation is not a function.   **2.** The relation is a function. The domain is the set of input values 1, 2, 3, and 4. The range is the set of output values 1, 4, 9, and 16.   **3.** The relation is a function. The domain is the set of input values 1, 2, 3, and 4. The range is the set of output values 4, 6, and 8.

4. $f(3) = 29; f(0) = 2; f(-2) = -16$
5. $f(3) = 5.5; f(0) = 4; f(-2) = 3$
6. $f(3) = -18; f(0) = 3; f(-2) = 17$
7. $f(t) = 340t; f(1.5) = 510$ miles
8. $f(t) = 380t; f(1.5) = 570$ miles

Interdisciplinary Application

1. $f(t) = 341.35t$ **2.** day 32: 10,923.2 nautical miles; day 61: 20,822.35 nautical miles

3.

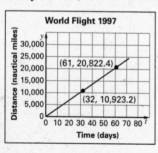

4. day 32: Massawa, Eritrea; day 61: Howland, USA

Challenge: Skills and Applications

1. 4, -4 **2.** no

3.

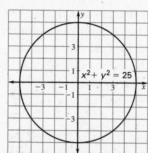

4. The relation is not a function because a vertical line can intersect the graph in more than one place.   **5.** *Sample answer:* $y \geq 0$ **6.** Accept reasonable answers. *Sample answer:* 0 to 4 seconds   **7.** Accept reasonable answers. *Sample answer:* 0 to 15 feet   **8.** *Sample answer:* all whole numbers of people from 0 through 30

9. *Sample answer:* all whole numbers of crafts from 0 through 90

10. *Sample answer:* (3, 5), (2, 1), (1, 2) (4, 6) (2, 2); 8, 3, 3, 10, 4

11. 36   **12.** 2, 3, 4, 5, 6, 7, 8, 9, 10, 11, 12

Review and Assessment

Review Games and Activities

SKIING AN INFINITE SLOPE;

MISSION IMPOSSIBLE

1. M   **2.** I   **3.** S   **4.** S; I   **5.** I; M   **6.** O; P

7. N; O   **8.** S   **9.** S   **10.** I   **11.** B   **12.** L

13. E   **14.** F

Test A

1. A: (2, 4); B: (3, -2); C: (-2, -3); D: (-2, 1)

2. A: (0, 1); B: (2, -3); C: (-4, -1); D: (-3, 3)

3. Quadrant IV   **4.** Quadrant III

5. Yes, (2, 7) is a solution of the equation.

6. No, (-3, 10) is not a solution of the equation.

7. 7   **8.** $\frac{3}{4}$ **9.** 4   **10.** 4

Review and Assesment *continued*

11.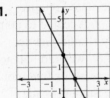

12. $\frac{1}{2}$ **13.** $-\frac{1}{3}$

14. 2 **15.** -1 **16.** $y = 5x$ **17.** $y = -4x$

18. $m = 2$; y-intercept: 5

19. $m = -3$; y-intercept: 5 **20.** $x = \frac{5}{4}$

21. $x = 1$ **22.** Yes, the lines are parallel.

23. No, the lines are not parallel.

24. Yes, the relation is a function.

25. Yes, the relation is a function.

Test B

1. No, it is not a solution.

2. Yes, it is a solution.

3. Sample table:

x	-1	0	1
y	$-\frac{9}{4}$	-2	$-\frac{7}{4}$

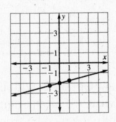

4. Sample table:

x	-1	0	1
y	$-\frac{7}{2}$	$-\frac{1}{2}$	$\frac{5}{2}$

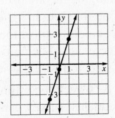

5. 2 **6.** $-\frac{2}{5}$ **7.** $\frac{7}{3}$ **8.** $-\frac{9}{4}$ **9.** $m = -4$

10. $m = \frac{5}{11}$ **11.** 5 **12.** 0

13. Rate of change = $30,000,000 per year

14. $y = -7x$ **15.** $y = 36x$

16. Yes, the two quantities have direct variation.

17. $y = \frac{3}{2}x - \frac{3}{4}$ **18.** $y = -\frac{5}{2}x + 5$

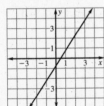

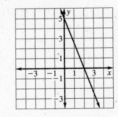

19. $x = \frac{1}{5}$ **20.** $x = 10$

21. No, the lines are not parallel.

22. Yes, the lines are parallel.

23. $f(3) = \frac{3}{2}; f(0) = 3; f(-2) = 4$

24. $h(3) = 20.5; h(0) = 4; h(-2) = -7$

25. $g(3) = -\frac{29}{8}; g(0) = -4; g(-2) = -\frac{17}{4}$

26. $k(3) = 2; k(0) = 14; k(-2) = 22$ **27.** $m = 3$

Test C

1. Yes, it is a solution.

2. No, it is not a solution.

3. Sample table:

x	0	1	3
y	$3\frac{1}{2}$	2	-1

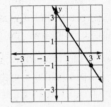

4. Sample table:

x	-1	0	1
y	-1	$-\frac{1}{5}$	$\frac{3}{5}$

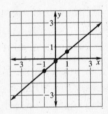

5. $2x + 5y = 400$ **6.** $-\frac{5}{13}$ **7.** -2

8. $-\frac{5}{2}$ **9.** $-\frac{13}{5}$ **10.** $m = \frac{11}{2}$

11. $m = \frac{25}{3}$ **12.** $\frac{24}{5}$ **13.** $-\frac{12}{7}$

14. Rate of change = 320,000 dollars per year

15. $x = 0.05y$

16. $y = -\frac{1}{9}x + \frac{1}{2}$ **17.** $y = \frac{3}{4}x + 2$

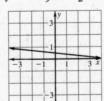

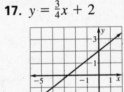

18. $x = -\frac{52}{9}$ **19.** $x = -57$

20. Yes, the lines are parallel.

21. No, the lines are not parallel.

22. $f(4) = 5.6; f(0) = -0.4; f(-3) = -4.9$

23. $h(4) = \frac{13}{6}; h(0) = \frac{2}{3}; h(-3) = -\frac{11}{24}$

24. $g(4) = -51; g(0) = 5; g(-3) = 47$

25. $k(4) = -16.8; k(0) = 3.2; k(-3) = 18.2$

26. $m = \frac{5}{3}$

SAT/ACT Chapter Test

1. A **2.** B **3.** D **4.** A **5.** D **6.** C **7.** A
8. B **9.** C

Alternative Assessment

1 a, b. Complete answers should address these points:

a. • Define *y*-intercept.
• Explain that a graph of $y = x + k$ is $|k|$ units above or below the graph of $y = x$.

b. • Define slope.
• Explain that a graph of $y = mx, m > 0$, slopes up from left to right and gets steeper as m increases.
• Explain that a graph of $y = mx, m < 0$, slopes down from left to right and gets steeper as m increases in absolute value.

2. a. increasing; 50 members per year; each year there were 50 new club members. **b.** 300; in 1985 there were 300 members. **c.** 1999
d. -200 members; *Sample answer:* This does not make sense because there can not be a negative number of people. The club probably was not in existence in 1975. **e.** 1800; *Sample answer:* The result may or may not be reasonable. The prediction is based on a model that ends in 2000, so it may not be very accurate for 2015.
f. Find the year when the club had no members; the *t*-intercept; 1979.

3. *Sample answer:* Both clubs had 300 members in 1985, because both graphs have the vertical intercept 300. The club in Exercise 2 is gaining members, because the rate of change is positive. The other club is losing members because the rate of change is negative. Therefore, the club in Exercise 2 is better at attracting members.

Project

1. Check that lists address the stated requirements. Note that students are not expected to come up with precise estimates. They should just do some "ballpark figuring" and find ways to balance expensive activities with inexpensive or free (contributed) ones. *Sample answer:* Student-made games/activities: bean toss, face painting, fishing game, balloon darts, fortune telling; Rented activities: dunk tank, basic inflatable play area; Food offerings; Food both (barbecued hot dogs and hamburgers, potato salad, cole slaw, chips, lemonade, and iced tea), bake sale of donated items, cotton candy. **2.** Option 1: $y = 0.25x - 300$, Option: $y = 0.50x - 500$

3. 800 tickets; $100 loss;

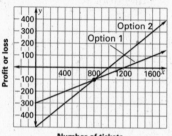

4. Option 2, with no admission, is better; *Sample answer:* For any number of tickets greater than 800, the income from Option 2 is greater since the line for Option 2 is above the line for Option 1. If you don't expect to sell more than 800 tickets there would be no point in having the carnival since you would lose money; 1000 tickets.

5. Check the estimates are reasonable based on students' survey data. Also check students' survey techniques to make sure they are using questions that seem clear and unbiased and that the sample is representative of the population for which the carnival is planned. **6.** Check equations and graphs. The equation for Option 2 will remain unchanged, but the *y*-intercept of the equation for Option 1 will vary depending on the number of people attending. The choice of option will depend on which one has the greater positive profit for the number of tickets the student expects to sell.

Review and Assesment *continued*

7. Using the original equations (based on an attendance of about 200 people), the carnival will make a profit if more than 1000 tickets could be sold. If a student made revisions in Exercise 6, check that the answer is consistent with the new equation, graph, and choice option. The carnival will make a profit if the number of tickets that can be sold is greater than x-intercept of the line for the selected pricing option.

Cumulative Review

1. 5 **2.** 3 **3.** -11 **4.** 2.1 **5.** $\frac{5}{4}$ **6.** $\frac{5}{9}$
7. 11 **8.** 4 **9.** 80 **10.** 3 **11.** 3 **12.** 20
13. $x + 6$ **14.** $15 - x$ **15.** $2x$ **16.** 2
17. 42 **18.** 17 **19.** -20 **20.** 6.5 **21.** 1.7
22. -4 **23.** 13 **24.** 2 **25.** $-\frac{1}{2}$ **26.** -1
27. $\frac{9}{4}$ **28.** $6z$ **29.** $-2b^3$ **30.** $-y^2$ **31.** $4|z^3|$
32. $-x^6$ **33.** x **34.** $\frac{1}{4}$ **35.** $\frac{3}{11}$ **36.** $\frac{17}{5}$
37. -2 **38.** 1 **39.** 617 **40.** about \$.38 per can **41.** \$.84 per gallon **42.** 7 dollars per hour
43. 1.25 liters per serving
44. Answers may vary. Sample: $(0, -1), (1, 1), (-1, -3)$ **45.** Answers may vary. Sample: $(0, 9), (1, 6), (-1, 12)$ **46.** Answers may vary. Sample: $\left(\frac{1}{4}, 0\right), \left(\frac{1}{4}, 1\right), \left(\frac{1}{4}, 2\right)$
47. Answers may vary. Sample: $(0, -1), (1, -1), (2, -1)$

48. Answers may vary. Sample:

Input, x	Output, y
-10	12
-5	7
0	2
5	-3
10	-8

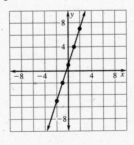

49. Answers may vary. Sample:

Input, x	Output, y
-2	-5
-1	-2
0	1
1	4
2	7

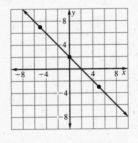

50. Answers may vary. Sample:

Input, x	Output, y
-10	9
-4	6
0	4
4	2
10	-1

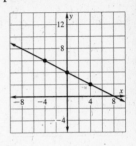

51. Answers may vary. Sample:

Input, x	Output, y
-2	-2
-1	-2
0	-2
1	-2
2	-2

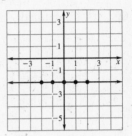

52. x-intercept $= -3$, y-intercept $= 3$

53. x-intercept $= \frac{9}{2}$, y-intercept $= 3$

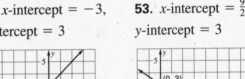

54. x-intercept $= \frac{1}{2}$, y-intercept $= -1$

55. x-intercept $= 7$, y-intercept $= -\frac{7}{4}$

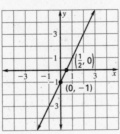
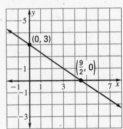

56. $m = -1$

57. $m = \frac{1}{3}$

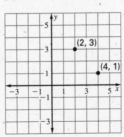
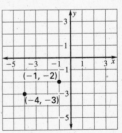

58. $m = 0$

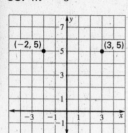

59. $m = \frac{8}{9}$

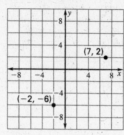

60. m is undefined

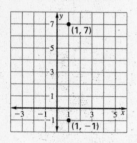

61. $m = -1$

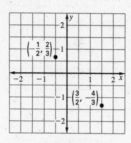

62. $y = 2$

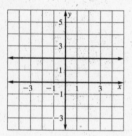

63. $y = \frac{1}{2}x - 2$

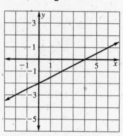

64. $y = -\frac{1}{5}x + \frac{1}{5}$

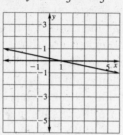

65. $y = x$

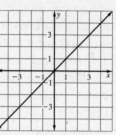

66. Yes, both lines have a slope of 2.

67. No, the slopes of the lines are different.

68. $y = 2x$ **69.** $y = 2x$

70. $y = \frac{1}{3}x$ **71.** $y = x$

72. $f(3) = 43; f(0) = -2; f(-1) = -17$

73. $f(3) = 12; f(0) = 0; f(-1) = -4$

74. $f(3) = -23; f(0) = 1; f(-1) = 9$

75. $f(3) = -\frac{3}{2}; f(0) = -3; f(-1) = -\frac{7}{2}$